Contents

Preface

The subject of polymer chemistry is an extremely important one, not only because it makes a fascinating study in itself but because it also plays a major part in the economy of most industrialised countries. This book is designed primarily for GCE A-level students, and also some teachers, who are interested in the topic of polymer chemistry and wish to broaden their knowledge of the subject. It is hoped that the book will be of particular help to those students who choose polymer chemistry as a special option for their particular board or syllabus. In addition, parts of the text may well be useful in some first year university courses and also in colleges which provide the TEC certificate and higher certificate courses.

The text contains an account of polymerisation techniques (namely addition and condensation polymerisation), an account of polymer structure and properties, and a short chapter on polymer characterisation by such methods as end-group analysis, viscometry, osmometry and light scattering. Two further chapters deal with the preparation and uses of common addition and condensation polymers, and the importance of naturally occurring high polymers. Finally, there is a list of suggested further reading together with some simple practical exercises.

Nomenclature, where practicable, follows the latest recommendations made by the Association for Science Education. I am grateful to Mr R. B. Jones (formerly of Unilever Research and Synthetic Resins Ltd) for advice on phenol–methanal and urea–methanal resins, and to Dr H. Block (Reader in Chemistry, University of Liverpool) for reading the manuscript and making many useful suggestions.

November 1981   *M. A. Cowd*

Polymer Chemistry

M. A. Cowd

BSc PhD CChem MRSC
Lecturer in Chemistry, North Wirral College of Technology

John Murray   Albemarle Street   London

MODERN CHEMISTRY BACKGROUND READERS

Edited by **J. G. Stark**
Head of Chemistry, Glasgow Academy

Other titles in the series:

Chemical Periodicity *D. G. Cooper*
Electrochemistry *P. D. Groves*
Inorganic Complexes *D. Nicholls*
The Shapes of Organic Molecules *N. G. Clark*

First published 1982
by John Murray (Publishers) Ltd.
50 Albemarle Street, London W1X 4BD
All rights reserved. Unauthorised duplication contravenes
applicable laws

Text set in 10/11 pt Linotron Times New Roman,
printed by photolithography, and bound in Great Britain by
J. W. Arrowsmith Ltd, Bristol BS3 2NT

British Library Cataloguing in Publication Data

Cowd, M. A.
Polymer Chemistry—(Modern chemistry
  background readers)
  1, Polymers and polymerization
  I. Title   II. Series
  547.7   QD381

0 7195 3961 7

Introduction 1

High polymers are molecules having high relative molecular masses. They can occur in nature (living matter, both animal and vegetable, consists largely of polymeric material) and they can also be synthesised in the laboratory. As will be seen later, chemists have acquired sufficient knowledge to allow them to tailor-make polymers for specific tasks or purposes, and it is this knowledge that has helped the polymer industry to expand so dramatically over the last forty years or so.

Whilst many naturally occurring polymers, such as cellulose, starch and proteins, have been known and used by man for centuries (for clothing and food), the polymer industry itself is a far more recent institution. Natural rubber was used in rubberised fabrics a few years before Goodyear's discovery of vulcanisation in 1839. Cellulose nitrate (produced by treating paper with nitric acid) was first produced industrially in 1870, phenolic resins in 1907 and poly(phenylethene) (polystyrene) in about 1930. Poly(ethene) (polyethylene or polythene) was first discovered in the ICI laboratories at Winnington, Cheshire, in 1933. Since this time, exciting and major breakthroughs have been made both in new polymeric systems and in the developments of existing ones, leading to the large-scale industrial output of polymers which exists today.

High polymers (sometimes called macromolecules) are large molecules built up by the repetition of small, simple chemical units; these repeating units are equivalent, or nearly equivalent, to the *monomer*, which is the starting material from which the polymer is made (see Table 1). Consequently, these molecules can often have very high relative molecular masses, e.g. some poly(phenylethene) samples have average relative molecular mass values of approximately

Table 1

Polymer	Monomer	Repeating unit
Poly(ethene)	$CH_2{=}CH_2$	$+CH_2-CH_2+$
Poly(chloroethene)	$CH_2{=}CHCl$	$+CH_2-CHCl+$
Cellulose	$C_6H_{12}O_6$	$+C_6H_{10}O_5+$

300 000—this is one of the reasons why high polymers can exhibit very different properties from those of low relative molecular mass materials of similar composition (other reasons will become evident as we progress through the text).

If the repetition of the repeating unit is *linear* (in much the same way as a chain is built up from its links), then the polymer molecules are often described as *chain molecules* or as *polymer chains* (Figure 1(a)). The length of the polymer chain can be described in terms of its *degree of polymerisation* (D.P.), which is the number of repeating units in the chain. Using poly(chloroethene) (polyvinyl chloride or PVC) as an example, a polymer having a D.P. of 1000 would have a relative molecular mass of $62.5 \times 1000 = 62\ 500$. However, polymer chains can also be *branched* (Figure 1(b)). In addition, separate linear or branched chains may be joined together along the chains by *cross-links* to give a *cross-linked polymer*. If cross-linking occurs to a high degree, three-dimensional cross-linked or *network polymers* can be formed (Figure 1(c)). Sometimes, cross-linking is carried out deliberately during manufacturing processes to modify the properties of a polymer, e.g. in the vulcanisation of rubber (see p. 87). Many polymeric systems rely almost entirely for their properties on the

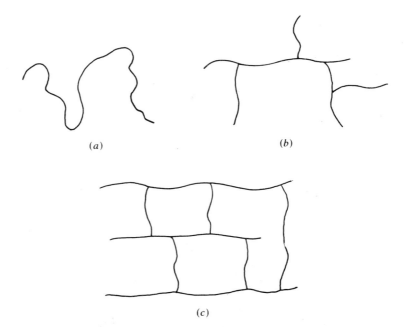

(a)   (b)

(c)

Figure 1 Schematic representation of (a) a randomly coiled linear chain, (b) a branched chain and (c) a network polymer.

development of three-dimensional networks, e.g. 'Bakelite', which is a phenol–methanal (phenol–formaldehyde) thermosetting resin (see p. 68). In such systems, extensive three-dimensional cross-linking occurs in the final stages of production, imparting rigidity and hardness on the polymer. If these final stages involve the use of heat, the polymer is classed as *thermosetting*, and the polymer is said to be *cured*. However, some polymer systems can be cured in the cold and are therefore termed *cold-setting*. Conversely, many linear polymers (containing few or no cross-links) can be softened and moulded on heating, and are described as *thermoplastic*.

Polymerisation

It was Dr W. H. Carothers, an American chemist, who classified polymerisation (i.e. the process of forming a high polymer) into two groups known as *addition polymerisation* and *condensation polymerisation*.

Addition polymerisation involves chain reactions (see p. 7), in which the chain carrier may be a free radical (a reactive species containing an unpaired electron) or an ion. A free radical is usually formed by the decomposition of a relatively unstable substance called an *initiator* (some examples of these are given in Chapter 2), which then initiates chain building. This process occurs very rapidly, often of the order of a second. Addition polymerisation is particularly common with double-bonded compounds, such as ethene (ethylene) and its derivatives.

Condensation polymerisation is regarded as analogous to condensation (or addition–elimination) reactions in low relative molecular mass materials. In such polymerisations, reaction occurs between two poly-functional molecules (a polyfunctional molecule is a molecule containing two or more groups which are capable of reaction) to give one larger polyfunctional molecule accompanied by the elimination of a small molecule, such as water. For example, consider the reaction between the bifunctional monomers ethane-1,2-diol (ethylene glycol) and benzene-1,4-dicarboxylic acid (terephthalic acid):

$$HOCH_2 \cdot CH_2OH + HOOC-\!\!\left\langle\bigcirc\right\rangle\!\!-COOH \longrightarrow$$

$$HOCH_2 \cdot CH_2OOC-\!\!\left\langle\bigcirc\right\rangle\!\!-COOH + H_2O$$

Note that the product is still bifunctional and so further reaction can now occur, producing a linear polymer, until one of the reactants is used up. It should also be noted that both branched and cross-linked polymers can be formed by both types of polymerisation (as will be seen in Chapter 2).

The term *copolymer* refers to a polymer made from polymerising two or more different suitable monomers together (thus offering a way of introducing variety into the structure of a polymer). Different types of copolymers exist. In a *random copolymer*, the different repeating units are arranged randomly in a chain; in an *alternating copolymer*, the different units alternate in a particular chain; in a *block copolymer*, blocks of one type of repeating unit alternate with blocks of another type; finally, in a *graft copolymer*, blocks of one type of repeating unit are attached or grafted to the backbone of a linear polymer containing the other type of repeating unit (Figure 2).

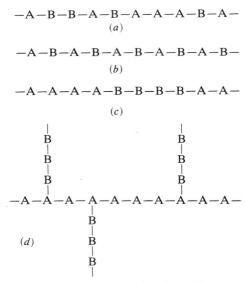

Figure 2 Diagrammatic representation of (a) random, (b) alternating, (c) block and (d) graft copolymers, where A and B represent the different repeating units.

Polymer properties

It must be appreciated that the process of chain growth during polymerisation is governed by random events. Therefore, different polymer chains within the same sample of polymer will have different lengths and therefore different relative molecular masses (M_r). Consequently, any experimental determination of the relative molecular mass, for a polymer sample, will only give an average value. Details of these average values, relative molecular mass distributions, and methods of determining M_r are outlined in Chapter 4.

Relative molecular mass is only one of the factors which affects polymer properties. Another major factor is the way in which the polymer chains are arranged within the polymer. X-ray studies have shown that within the bulk polymer there are regions where chains are arranged in a highly ordered manner—these regions are referred to as *crystalline regions* or *crystallites*. Between these ordered regions there are *amorphous regions* where the chains are tangled or in a state of disorder (Figure 3). The crystalline regions can be formed when chains are able to approach each other closely enough for strong interchain forces to operate. Several factors will determine whether this can happen (and these are discussed in Chapter 3), but it should be easy to see that linear chains will be able to approach each other more closely than branched chains of the same polymer.

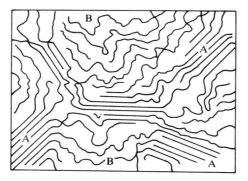

Figure 3 The crystalline–amorphous structure of polymers, where A represents a crystalline region and B an amorphous region.

The degree of crystallinity has a great effect on the properties (and therefore uses) of the polymer. For example, polymers with symmetrical repeating units and high interchain forces can be used to form fibres having high crystallinity and tensile strength. Plastics, or resins, however, have a smaller degree of crystallinity and, unless they are highly cross-linked, may be softened and moulded at higher temperatures. Elastomers, or rubbers, are highly extensible (may be stretched) and elastic, having a very low crystallinity.

The industrial situation

The production of raw polymers and their conversion to finished and usable articles is the basis of the polymer industry. However, many other industries are involved, e.g. the mechanical and chemical engineering companies who manufacture the necessary plant and equipment needed for polymer production and conversion. In addition, the use of polymeric components in other industries, e.g. the motor and electrical industries, is enormous. Consequently, the polymer industry can be regarded as one of the basic industries in the economy of any industrialised country.

Despite the oil crisis of 1973 (which meant an escalation in prices of rubbers and plastics) and warnings of over-production by plastics producers, world output of polymers is still high. Table 2 summarises the position for several major countries, using some common addition polymers as examples.

Table 2 Output of some common polymers (thousands of tonnes).

Canada

Polymer	1976	1977	1978	1979
Poly(ethene)	300.9	345.0	477.5	591.0
Poly(phenylethene)	75.9	90.5	98.7	119.7
Poly(chloroethene)	96.8	99.7	124.7	152.4

Japan

Polymer	1976	1977	1978	1979
Poly(ethene)	1392	1467	1767	2165
Poly(phenylethene)	876	900	1032	1227
Poly(chloroethene)	1044	1030	1204	1583

United Kingdom

Polymer	1976	1977	1978	1979
Poly(ethene)	473.1	487.2	427.3	459.9
Poly(phenylethene)	238.6	228.6	184.2	220.4
Poly(chloroethene)	415.9	385.1	409.1	425.0

West Germany

Polymer	1976	1977	1978	1979
Poly(ethene)	1466.9	1431.1	1518.6	1587.6
Poly(chloroethene)	962.9	894.9	1003.8	1085.9

United States of America

Polymer	1976	1977	1978	1979
Poly(ethene)	4054.2	4591.7	5130.0	5807.4
Poly(chloroethene) and copolymers	2139.1	2382.7	2595.9	2775.5

In terms of research, there has been a change in emphasis from polymer chemistry to the physics and engineering aspects of established systems, e.g. relationships between structure and properties is a key area (i.e. the problem of 'tailoring' a polymer for a particular task), and also the electrical properties of polymers. Moves to establish entirely new high tonnage polymers from new monomers have declined considerably.

Polymerisation 2

As has already been mentioned (p. 3), polymerisation can be classified into two main groups known as addition and condensation polymerisation.

Addition polymerisation

This type of polymerisation involves chain reactions, where the chain carrier can be a reactive species containing an unpaired electron (called a *free radical*) or it can be an ion. The more important polymers produced by this method include those formed from ethene derivatives of the type $CH_2{=}CHX$ or $CH_2{=}CXY$.

The basic features of a chain reaction can be shown by the reaction between hydrogen and chlorine. If chlorine atoms (usually produced by exposing molecular chlorine to light) are introduced into a mixture of hydrogen and chlorine, two rapid reactions occur (equations 2.2 and 2.3); this situation can be represented as follows (where a dot denotes an unpaired electron):

$$Cl_2 \xrightarrow{\text{u.v.}} 2Cl\cdot \tag{2.1}$$

then

$$H_2 + Cl\cdot \rightarrow HCl + H\cdot \tag{2.2}$$

$$H\cdot + Cl_2 \rightarrow HCl + Cl\cdot \tag{2.3}$$

As a result of this chain reaction, a molecule of chlorine and one of hydrogen have combined together to form two molecules of hydrogen chloride, but note that a chlorine atom has been regenerated. This chlorine atom (or free radical) can initiate the further combination of other hydrogen and chlorine molecules. In other words, the original introduction of a few chlorine atoms has therefore initiated a chain of consecutive reactions which carry on repeating until one (or both) of the reactants is used up. This series of successive processes is known as a *chain reaction*, and the hydrogen and chlorine atoms (in equations 2.2 and 2.3) are known as the *chain carriers*. The chains are said to be *initiated* by 2.1 and *propagated* by 2.2 and 2.3. The chains are not of unlimited length—*termination*, which will remove chain carriers, can occur by the following processes:

$$Cl\cdot + Cl\cdot \ \rightarrow \ Cl_2 \tag{2.4}$$

$$H\cdot + Cl\cdot \ \rightarrow \ HCl \tag{2.5}$$

The propagation and termination processes are, generally speaking, extremely fast (sometimes even explosive). However, the whole reaction may take considerable time for completion since research has shown that a chain reaction occurs in a series of short bursts of activity or 'spurts'.

Addition polymerisation can also be split into these three stages of initiation, propagation and termination (see below). Since chain carriers can be free radicals or ions, addition polymerisation can therefore be classified into (i) free radical polymerisation and (ii) ionic (cationic and anionic) polymerisation.

(i) *Free radical polymerisation*

This is probably the most common and important type of addition polymerisation. Free radicals are usually formed by the decomposition of a relatively unstable material (using heat or sometimes light) called an *initiator*. Examples of initiators include peroxide compounds, such as di(benzoyl) peroxide (benzoyl peroxide):

and various azo compounds; the one below is commonly referred to as AZDN (from its old name azodiisobutyronitrile):

Initiation can be regarded as (a) the decomposition of the initiator and (b) the addition of a monomer molecule to one of the resulting free radicals. If we represent a free radical (formed from the initiator, I) as $R\cdot$ and the monomer molecule as $CH_2=CHX$, then the initiation process can be represented by:

$$I \rightarrow 2R\cdot \qquad (2.6)$$

then

$$R\cdot \ + \ CH_2=CHX \rightarrow R-CH_2-\dot{C}HX \qquad (2.7)$$

Propagation is then the further addition of monomer molecules to the free radical formed in 2.7:

$$R-CH_2-\dot{C}HX \ + \ CH_2=CHX \rightarrow$$

$$R-CH_2-CHX-CH_2-\dot{C}HX \quad \text{and so on.}$$

Termination can occur either by *combination*:

$$\text{\textasciitilde}CH_2-\dot{C}HX \ + \ \dot{C}HX-CH_2\text{\textasciitilde} \rightarrow$$

$$\text{\textasciitilde}CH_2-CHX-CHX-CH_2\text{\textasciitilde}$$

or by *disproportionation*:

$$\text{\textasciitilde}CH_2-\dot{C}HX \ + \ \overset{H}{\underset{}{\dot{C}HX-CH}}\text{\textasciitilde} \rightarrow$$

$$\text{\textasciitilde}CH_2-CH_2X+CHX=CH\text{\textasciitilde}$$

Disproportionation involves the *transfer* of the penultimate or β-hydrogen atom from one radical to another, producing two inactive polymer molecules, one of which has a terminal double bond whilst the other is saturated.

Whilst the three steps of initiation, propagation and termination are both necessary and sufficient for chain polymerisation, other processes, such as *chain transfer*, can occur during polymerisation. Here, an atom (usually a hydrogen or halogen atom) is transferred from a molecule (called the *transfer agent*) to the radical. This transfer agent could be a molecule of solvent, initiator (not common, since the concentration of initiator is usually low), monomer or polymer. For example, consider the situation when tetrachloromethane (carbon tetrachloride) is acting as a solvent in the polymerisation of our monomer, $CH_2=CHX$:

$$\text{\textasciitilde}CH_2-\overset{H}{\underset{X}{\overset{|}{C}}}\cdot+CCl_4 \rightarrow \text{\textasciitilde}CH_2-\overset{H}{\underset{X}{\overset{|}{C}}}-Cl+\cdot CCl_3 \qquad (2.8)$$

Here, a chlorine atom has been transferred to the growing radical leaving an inactive polymer molecule and a $\cdot CCl_3$ radical. The latter can now initiate the growth of another polymer chain:

$$\cdot CCl_3 + CH_2{=}CHX \rightarrow CCl_3{-}CH_2{-}\dot{C}HX \quad \text{and so on.} \tag{2.9}$$

Clearly, the more frequently chain transfer occurs the shorter the chain will be (on average) since its growth has been interrupted—hence the degree of polymerisation (D.P.—see p. 2) is reduced and therefore the relative molecular mass (M_r). Transfer, in fact, can have special industrial importance. For example, polymerisations sometimes yield polymers with such high values of M_r that difficulties are experienced in moulding them. Hence, the addition of a transfer agent will decrease the degree of polymerisation to a level which will enable this problem to be overcome.

If the transfer agent is a polymer molecule itself, branching (see p. 2) can occur:

$$\text{\small www}CH_2{-}\dot{C}HX + \text{\small www}CH_2{-}CHX\text{\small www} \rightarrow \text{\small www}CH_2{-}CH_2X + \text{\small www}CH_2{-}\dot{C}X\text{\small www}$$

Then, on further reaction with monomer:

$$\text{\small www}CH_2{-}\dot{C}X\text{\small www} + CH_2{=}CHX \rightarrow \text{\small www}CH_2{-}\underset{\underset{\cdot CHX}{\overset{|}{CH_2}}}{\overset{|}{C}}X\text{\small www} \quad \text{and so on.}$$

Note also that the combination of two growing branches on two different polymer chains would result in a cross-link being formed (see p. 2):

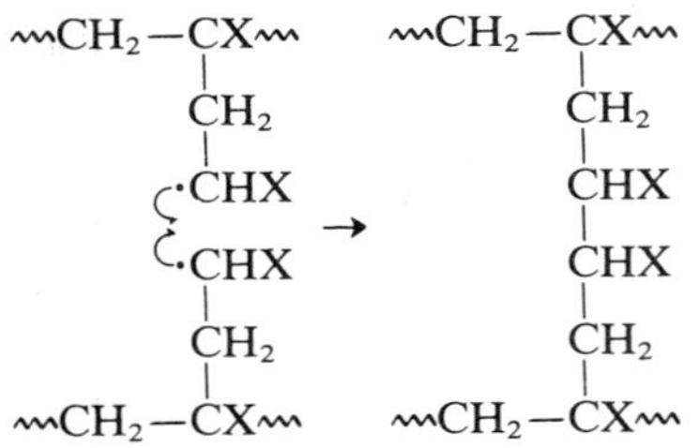

Both the rate and degree of polymerisation can be controlled by the use of substances known as *retarders* and *inhibitors*. Inhibitors react with radicals as soon as they are produced. Therefore, polymerisation cannot occur until all the inhibitor has been used up (the period of time during which polymerisation is inhibited is called the *induction period*). Once the inhibitor has been used up, the rate of polymerisation then approaches the same value as that for normal uninhibited polymerisation. Quinones can act as inhibitors for many polymerisation systems; it is thought that they react with the free radicals, as they are produced, to form resonance-stabilised radicals which are unable to initiate polymerisation. An important application of inhibition is in the stabilisation of monomers. Storage of monomers in the pure state can be dangerous since if free radicals are formed (from peroxides produced by oxidation, for example), polymerisation could begin and possibly accelerate to a dangerous level. The addition of small quantities of inhibitor to the monomer prevents this. The inhibitor is then removed, by distillation of monomer, for example, before the monomer is used.

Retarders are less reactive then inhibitors and compete with the monomer for free radicals; hence, both the rate and degree of polymerisation are reduced. The simplest type of retarder is a free radical which is too unreactive to initiate a polymer chain itself but can interrupt polymer growth by, for example, combining with the propagating radical. Sometimes, an atom is transferred from a retarder molecule to the propagating radical to produce an inactive polymer molecule, just as in equation 2.8; this time, however, a new inactive radical is formed which cannot re-initiate polymerisation as in the normal transfer process. Alternatively, the retarder molecule may simply add on to the propagating radical to form a new radical which is slow to react with further monomer. Oxygen is a strong retarder of many free radical polymerisations, adding on to the propagating radical as described:

$$\text{\textasciitilde\textasciitilde\textasciitilde} CH_2 - \overset{\cdot}{C}HX + O_2 \rightarrow \text{\textasciitilde\textasciitilde\textasciitilde} CH_2 - CHX - O - \overset{\cdot}{O}$$

Hence, in oxygen-sensitive polymerisations, the oxygen has to be removed prior to polymerisation being carried out.

(ii) Ionic polymerisation

Addition polymerisation can occur by mechanisms other than those involving free radicals. For example, the chain carriers can be carbonium ions (cationic polymerisation) or carbanions (anionic polymerisation). Some of these mechanisms are now discussed.

(a) *Cationic polymerisation.* In cationic polymerisations of our monomer, $CH_2=CHX$, the chain carriers are carbonium ions. Typical catalysts for this type of polymerisation are Lewis acids (electron pair acceptors) and Friedel–Crafts catalysts—examples include $AlCl_3$, $AlBr_3$, BF_3, $TiCl_4$, $SnCl_4$, H_2SO_4 and other strong acids. Unlike free radical polymerisations which, generally speaking, occur at elevated temperatures, cationic polymerisations proceed best at low temperatures, e.g. 2-methylpropene (isobutylene) polymerises extremely rapidly in the presence of BF_3 or $AlCl_3$ at $-100\,°C$. Also, the influence of solvent is important (see later) because ionic mechanisms involve charged species whereas free radicals are generally neutral. Cationic polymerisations often occur with monomers containing electron-releasing groups.

In acid-catalysed polymerisations, initiation can be represented as follows (where HA is an acid molecule such as hydrochloric, sulphuric or chloric(VII) acid (perchloric acid)):

$$HA + H_2C{=}CHX \rightarrow H_3C-\underset{\diagdown X}{\overset{\diagup H}{C_+}} \quad + A^- \tag{2.10}$$

In other words, proton transfer from the acid to the monomer, producing a carbonium ion, has occurred. Propagation is then the further addition of monomer to the carbonium ion produced in 2.10, much as in the free radical scheme which we have already considered:

$$H_3C-\underset{\diagdown X}{\overset{\diagup H}{C_+}} \quad + H_2C{=}CHX \rightarrow H_3C-\underset{\underset{H}{|}}{\overset{\overset{H}{|}}{C}}-\underset{\underset{H}{|}}{\overset{\overset{H}{|}}{C}}-\underset{\diagdown X}{\overset{\diagup H}{C_+}} \quad \text{and so on.}$$

Termination of the chain will be considered presently.

In the case of Friedel–Crafts catalysts, since they contain no hydrogen, proton transfer, obviously, cannot occur as in 2.10. It is found, therefore, that the presence of a *co-catalyst* is necessary for polymerisation to proceed, the commonest co-catalyst being water. For example, it has been shown that anhydrous Friedel–Crafts catalysts do not initiate polymerisation; however, on introducing a trace of water, polymerisation takes place. Using BF_3 as our example, the mechanism can be represented as follows:

$$BF_3 + H_2O \rightarrow BF_3.H_2O \tag{2.11}$$

$$BF_3.H_2O + H_2C{=}C\underset{\diagdown X}{\overset{\diagup H}{}} \rightarrow H_3C-\underset{\diagdown X}{\overset{\diagup H}{C_+}} \quad + [BF_3OH]^- \tag{2.12}$$

In other words, proton transfer to the monomer has now been made possible. Propagation is then the further addition of monomer to the carbonium ion produced in 2.12.

Termination of the chain can occur by several processes. The simplest and most obvious process is the combination of the carbonium ion and its associated anion (called the counter ion):

$$\text{ⱮCH}_2-\underset{\diagdown X}{\overset{\diagup H}{C_+}} \quad + A^- \rightarrow \text{ⱮCH}_2-\underset{\underset{X}{|}}{\overset{\overset{H}{|}}{C}}-A$$

Alternatively, termination can take place by the rearrangement of the ion pair to give unsaturation in the polymer chain and the original acid or complex. For example, in the polymerisation of 2-methylpropene with BF_3, the following proton transfer can occur:

$$\text{wwCH}_2-\overset{\displaystyle CH_3}{\underset{\displaystyle CH_3}{C^+}}\quad [BF_3OH]^- \;\rightarrow\; \text{wwCH}_2-\overset{\displaystyle CH_2}{\underset{\displaystyle CH_3}{C}} \quad +H^+[BF_3OH]^- \;\;(\text{or } BF_3.H_2O)$$

Termination can also occur via chain transfer to monomer. Using the above system as our example:

$$\text{wwCH}_2-\overset{\displaystyle CH_3}{\underset{\displaystyle CH_3}{C^+}}\quad [BF_3OH]^- + H_2C=\overset{\displaystyle CH_3}{\underset{\displaystyle CH_3}{C}}$$

$$\downarrow$$

$$\text{wwCH}_2-\overset{\displaystyle CH_2}{\underset{\displaystyle CH_3}{C}}\quad + H_3C-\overset{\displaystyle CH_3}{\underset{\displaystyle CH_3}{C^+}}\quad [BF_3OH]^-$$

Finally, it should be noted that the nature of the solvent is very important in cationic polymerisations. For example, it is observed with some systems that no polymerisation occurs if the solvent used is non-polar. Instead, the carbonium ion, when first produced, reacts with its counter ion giving a reaction similar to the normal electrophilic addition of a compound, HA, to a double-bonded compound, e.g. hydrogen halide with an alkene. If, however, the same reaction is repeated using a polar solvent, polymerisation occurs. This is because the carbonium ion becomes stabilised by solvation and so the counter ion is kept away to some degree. Addition of (neutral) monomer to the carbonium ion can now take place to produce polymer.

(b) *Anionic polymerisation.* In anionic polymerisations of our monomer, $CH_2=CHX$, the chain carriers are carbanions. Monomers containing electronegative substituents, such as propenenitrile (acrylonitrile), methyl 2-methylpropenoate (methyl methacrylate), and even phenylethene (styrene), are particularly susceptible to this type of polymerisation. Like cationic polymerisations, reaction proceeds best at low temperatures. Suitable catalysts include the alkali metals, and alkali metal alkyls, aryls and amides. In fact, one of the very early polymerisations used in industry was the production of synthetic rubber, in Germany and Russia, from buta-1,3-diene (butadiene) catalysed by alkali metals.

As an example of anionic polymerisation, let us consider an alkali metal amide, such as potassium amide (or potassamide), KNH_2, in liquid ammonia acting on the monomer, $CH_2{=}CHX$. In liquid ammonia, potassium amide is strongly ionised—initiation can be represented by:

$$H_2N^- + H_2C{=}C\overset{H}{\underset{X}{\diagdown}} \rightarrow H_2N-\overset{H}{\underset{H}{\overset{|}{C}}}-\overset{H}{\underset{X}{\overset{|}{C}}}{:}^- \qquad (2.13)$$

The positive counter ion in this case is obviously the K^+ ion. Propagation is then the addition of further monomer to the carbanion produced in 2.13:

$$H_2N-\overset{H}{\underset{H}{\overset{|}{C}}}-\overset{H}{\underset{X}{\overset{|}{C}}}{:}^- + H_2C{=}C\overset{H}{\underset{X}{\diagdown}} \rightarrow H_2N-\overset{H}{\underset{H}{\overset{|}{C}}}-\overset{H}{\underset{X}{\overset{|}{C}}}-\overset{H}{\underset{H}{\overset{|}{C}}}-\overset{H}{\underset{X}{\overset{|}{C}}}{:}^- \quad \text{and so on.}$$

However, the termination process is not so clear to see as in the previous addition polymerisations we have discussed. Combination of the anionic chain with its metal counter ion, or loss of a hydride ion, H^-, from the anionic chain to form a metal hydride (e.g. K^+H^-) and an inactive chain containing an unsaturated end group, are both considered unlikely possibilities. In fact, provided that inert solvent and pure reactants are used, polymerisation only stops when all the available monomer has been used up. Even then, the active centres (or carbanions) are not destroyed, and if more monomer were added, polymerisation could recommence. Such polymers are known as 'living polymers'. However, traces of water, carbon dioxide, alcohols and other materials will terminate the chains:

$$\sim\!\!\text{C}-\overset{H}{\underset{H}{\overset{|}{C}}}{:}^-M^+ + H_2O \rightarrow \sim\!\!\text{C}-\overset{H}{\underset{X}{\overset{|}{C}}}-H + M^+OH^-$$

and

$$\sim\!\!\text{C}-\overset{H}{\underset{H}{\overset{|}{C}}}{:}^-M^+ + O{=}C{=}O \rightarrow \sim\!\!\text{C}-\overset{H}{\underset{X}{\overset{|}{C}}}-\overset{O}{\overset{\|}{C}}-O^-M^+$$

Finally, not all catalysts used for addition polymerisations can be easily fitted into the previous categories discussed. A particularly important type of catalyst is the so-called Ziegler–Natta catalyst,* discovered by Ziegler in 1953. He used this system for polymerising ethene (ethylene); later, Natta extended its use (in 1955) to the polymerisation of propene (propylene) and other unsaturated

---

* More commonly referred to as just a Ziegler catalyst.

monomers. These catalyst systems can be prepared by mixing an alkyl or aryl of an element from groups I–III of the Periodic Table with a halide of a transition element. For example, triethylaluminium (aluminium triethyl), $Al(C_2H_5)_3$, when added to titanium(IV) chloride (titanium tetrachloride) in hexane solution produces a brown-black precipitate which catalyses the polymerisation of ethene at low pressures. Prior to this catalyst being used by Ziegler, poly(ethene) (polyethylene or polythene) had been produced by free radical polymerisation at pressures between 1000 and 3000 atmospheres and temperatures up to 250 °C; this method had resulted in branched polymers being formed. In contrast, Ziegler's method produced poly(ethene) which was essentially linear. It was also found that by using Ziegler–Natta catalysts, polymers, such as poly(propene) (polypropylene), could be produced which had high degrees of regularity in the configurations of their chains (this is discussed further on p. 26). Polymerisation reactions of this type involving substituted ethenes are said to be *stereospecific*. Whilst the exact mechanism of these reactions is still in dispute, an account of the evidence accumulated so far is beyond the scope of this text. It can be said, however, that the monomer is adsorbed in some way by the catalyst; consequently, some steric control of the propagation reaction can then take place during polymerisation.

Condensation polymerisation

This type of polymerisation (see p. 3) involves the joining of small molecules, to produce larger ones, using the common condensation (or addition–elimination) reactions of organic chemistry. For example, if ethanol (ethyl alcohol) and ethanoic acid (acetic acid) are warmed together in the presence of a small amount of concentrated sulphuric acid, an ester (ethyl ethanoate (ethyl acetate)) is produced accompanied by the elimination of water:

$$CH_3.COOH + C_2H_5OH \rightleftharpoons CH_3.COOC_2H_5 + H_2O \qquad (2.14)$$

Note, however, that the reaction stops at this stage since there are no functional groups remaining to react (in this example, the functional groups are the —COOH and —OH groups). If, however, each reacting molecule contains two or more functional groups, then the reaction can proceed further. For example, consider the reaction between the two monomers hexanedioic acid (adipic acid) and ethane-1,2-diol (ethylene glycol):

$$HOOC.(CH_2)_4.COOH + HOCH_2.CH_2OH \rightarrow$$

$$HOOC.(CH_2)_4.COOCH_2.CH_2OH + H_2O$$

It can be seen that in this case the product still contains two functional groups, so further reaction with monomer can now occur at either end of the molecule to produce a larger molecule:

$$HOCH_2.CH_2OH + HOOC.(CH_2)_4.COOCH_2.CH_2OH$$
$$\downarrow$$
$$HOCH_2.CH_2OOC.(CH_2)_4.COOCH_2.CH_2OH + H_2O$$

These polymerisations, therefore, nearly always proceed in a stepwise manner by reaction between pairs of functional groups; hence, dimer, trimer, tetramer and eventually polymer is produced (the polymer, in this case, contains the repeating unit $-OCH_2.CH_2OOC.(CH_2)_4.CO-$). Consequently, the relative molecular mass rises steadily during reaction and reaction times are long if high relative molecular mass polymer is required. This is in contrast to chain or addition polymerisation where high relative molecular mass polymer can be formed virtually immediately. Notice also that reaction between two growing molecules can also occur:

$$\sim\!\!\sim CH_2OH + HOOC\!\sim\!\!\sim$$
$$\downarrow$$
$$\sim\!\!\sim CH_2OOC\!\sim\!\!\sim + H_2O$$

In fact, it has been shown that the reactivity of a functional group at the end of a polymer molecule is similar to that of the same functional group in a molecule of monomer—in other words, the reactivity of a functional group is independent of the size of the molecule to which it is attached.

Condensation polymerisation usually involves the loss of water or other small molecules (but not always—see below) during polymerisation. The reader will remember that in esterification reactions of the type shown in equation 2.14, the amount of product is determined by the position of equilibrium—the equilibrium can be shifted forwards or to the right, hence increasing the yield of ester, by the removal of water as it is formed. Similar equilibria are also set up at each step of reaction during polymerisations of systems such as our hexanedioic acid–ethane-1,2-diol system. Consequently, removal of eliminated molecules, such as water, can be important in shifting the equilibrium forwards, hence encouraging further reaction, and therefore producing higher relative molecular mass polymer.

At this stage, it should be mentioned that the elimination of a small molecule at each step of reaction is not essential for the polymerisation to be classed as condensation polymerisation; one example of such a reaction is the production of a polyurethane from a diol (glycol) and a diisocyanate—see Chapter 5—(however, some authors do refer to this type of polymerisation as 'polymerisation through functional groups', keeping the term 'condensation' just for polymerisations involving the elimination of small molecules):

$$OCN-R-NCO+HO-R'-OH$$
$$\downarrow$$
$$OCN-R-NH-CO-O-R'-OH$$
$$\downarrow \text{further reaction with monomers}$$
$$+CO-NH-R-NH-CO-O-R'-O+_n$$

Here, reaction between functional groups involves the transfer of hydrogen from the hydroxyl group to the nitrogen atom of the $-NCO$ group. Since the polymer chain now contains the $-NH-CO-O-$ or urethane group, the polymer is said to be a polyurethane.

Up to now we have only looked at bifunctional monomers, in our examples, which have led to the production of linear polymers. However, these monomers can also lead to ring structures being formed, as shown below:

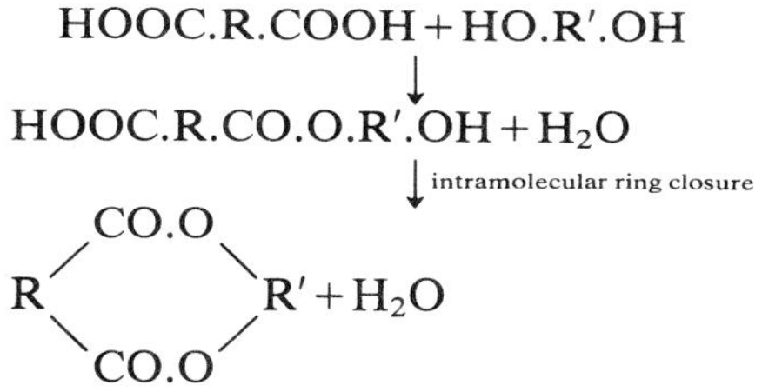

$$HOOC.R.COOH+HO.R'.OH$$
$$\downarrow$$
$$HOOC.R.CO.O.R'.OH+H_2O$$
$$\downarrow \text{intramolecular ring closure}$$

$$R\begin{array}{c} /CO.O\backslash \\ \\ \backslash CO.O/ \end{array}R'+H_2O$$

This ring formation is clearly an alternative to the normal (linear) chain growth discussed previously. Factors which determine whether ring structures will be formed include the flexibility of the groups involved and the size of the ring which may be formed. Some ring structures are more strained or unstable than others and, obviously, it is more likely that a linear polymer would be formed rather than strained or unstable rings. In addition to linear chains and cyclic products, branched and network polymers can be formed if one or both of the monomers has more than two functional groups. For example, consider the reaction between propane-1,2,3-triol (glycerol) and a dibasic acid. Since the triol possesses three functional groups, branched structures can be formed initially:

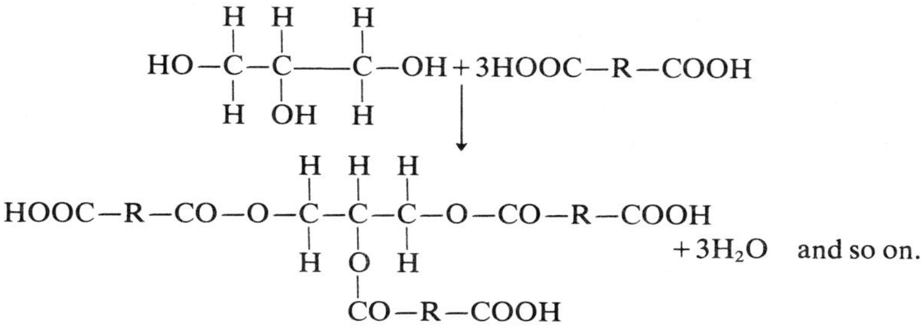

$$HO-\overset{\overset{\displaystyle H}{|}}{\underset{\underset{\displaystyle H}{|}}{C}}-\overset{\overset{\displaystyle H}{|}}{\underset{\underset{\displaystyle OH}{|}}{C}}-\overset{\overset{\displaystyle H}{|}}{\underset{\underset{\displaystyle H}{|}}{C}}-OH+3HOOC-R-COOH$$
$$\downarrow$$
$$HOOC-R-CO-O-\overset{\overset{\displaystyle H}{|}}{\underset{\underset{\displaystyle H}{|}}{C}}-\overset{\overset{\displaystyle H}{|}}{\underset{\underset{\displaystyle O}{|}}{C}}-\overset{\overset{\displaystyle H}{|}}{\underset{\underset{\displaystyle H}{|}}{C}}-O-CO-R-COOH$$
$$\underset{\displaystyle CO-R-COOH}{|}$$

$+3H_2O$ and so on.

As reaction continues and the size of the branched polymer molecules increases, there is an increased probability that condensation will occur in such a way as to link one polymer molecule with another—in other words, cross-linking occurs:

$$
\begin{matrix}
\text{\large\textasciitilde\textasciitilde\textasciitilde\textasciitilde} & \text{\large\textasciitilde\textasciitilde\textasciitilde\textasciitilde} \\
| & | \\
\text{COOH} & \text{C=O} \\
+ & \quad\quad | \\
\text{OH} \quad\longrightarrow & \text{O} \quad\quad +H_2O \\
| & | \\
\text{\large\textasciitilde\textasciitilde\textasciitilde\textasciitilde} & \text{\large\textasciitilde\textasciitilde\textasciitilde\textasciitilde}
\end{matrix}
$$

At some stage in the reaction, therefore, extremely large network structures (see p. 2) are formed, accompanied by a sudden change in polymer properties. For example, the reaction mixture changes from a viscous liquid to a gel; hence, the point at which the polymer network is formed throughout the polymerising system, or when gelation occurs, is called the *gel point*. Also, the polymer is no longer soluble in solvents which would dissolve the polymer before the gel point was reached (in fact, the gel will not dissolve in any solvent which does not attack it chemically). The industrial uses of three-dimensional networks are discussed later in Chapter 5.

Finally, the reader will have noticed the absence of a termination reaction in condensation polymerisation; polymerisation continues until, ideally, no more functional groups remain available for reaction. However, reaction and hence degree of polymerisation can be controlled by varying the time for reaction (see p. 21) and by adjusting the temperature, i.e. reaction can be stopped by cooling at the desired point, but polymerisation can restart if the temperature is subsequently raised. A more permanent way of stopping reaction is to use an 'end-stopper'. For example, addition of a small amount of ethanoic acid to the polymerising system is used to stabilise the relative molecular mass of nylon polymers (see p. 62). Since ethanoic acid is monofunctional, once it has reacted with the ends of the growing chains, no further reaction can occur. Hence, the degree of polymerisation (and therefore the relative molecular mass) of the resulting polymer can be controlled. A similar effect can be achieved by using an excess of one of the monomers. For example, if in a mixture of the two monomers   —A— and —B— an excess of —A— is used, then as —B— is used up the ends of the chain will consist of molecules of —A—, i.e. there will be an accumulation of chains such as   —A\textasciitilde\textasciitilde\textasciitilde\textasciitilde A—. Consequently, no further condensation can occur, and the degree of polymerisation will again be limited.

Copolymerisation

The simplest types of polymers are *homopolymers* where the repeating units are all of the same type or structure—in other words, homopolymers can be given a general formula of the type $X(A)_n Y$ where X and Y are the end or terminal groups (which we have already discussed) and A represents the repeating unit. Hence, polymers such as poly(ethene) and poly(chloroethene) (polyvinyl chloride or PVC) are both examples of homopolymers. If, however, two or more suitable monomers are polymerised together to give polymers containing more than one type of structural unit, a *copolymer* can be formed. In this case, we can write the general formula $X(A)_n(B)_m(C)_l \cdots Y$ where A, B, C, etc. represent the various structural units depending on the different monomers used. The various types of copolymers which can be made have already been outlined, and their general structures are shown in Figure 2 (p. 4). If monomers A and B are reacted together to form a copolymer, then the copolymer often shows very different properties from those of a physical mixture of the separate homopolymers of A and B. Sometimes, the good qualities of each homopolymer can be combined or retained in the copolymer, and this is one of the obvious advantages which copolymerisation offers. We will now discuss some of the ways in which copolymers can be produced.

Random copolymers are prepared by polymerising the appropriate mixture of monomers; examples include chloroethene–ethenyl ethanoate (vinyl chloride–vinyl acetate) and phenylethene–buta-1,3-diene copolymers. In the case of the chloroethene–ethenyl ethanoate copolymer, the presence of the ethenyl ethanoate increases solubility and improves moulding characteristics (by improving flow properties) as compared to the homopolymer of chloroethene. The properties of a copolymer produced from the monomers, A and B, will clearly depend on the distribution of A and B units in the chains of our copolymer. This distribution is not necessarily the same as the ratio of the concentrations of A to B in the original monomer mixture. Generally speaking, if two monomers A and B react to form a copolymer, and A is the more reactive monomer, then the copolymer formed in the early stages of polymerisation will be richer in A than in B. In the later stages of reaction, as the concentration of monomer A becomes low, the copolymer formed becomes richer in B. This problem of the copolymer composition changing during polymerisation may be reduced by adding the more reactive monomer gradually. The question of monomer reactivity can be treated quantitatively by studying the kinetics of copolymerisation, but this will not be dealt with here.

There are not many examples of truly alternating copolymers, but the best known example is the product obtained by the free radical copolymerisation of butenedioic anhydride (maleic anhydride) and phenylethene in approximately equimolar proportions. The structure of the chain can be written as:

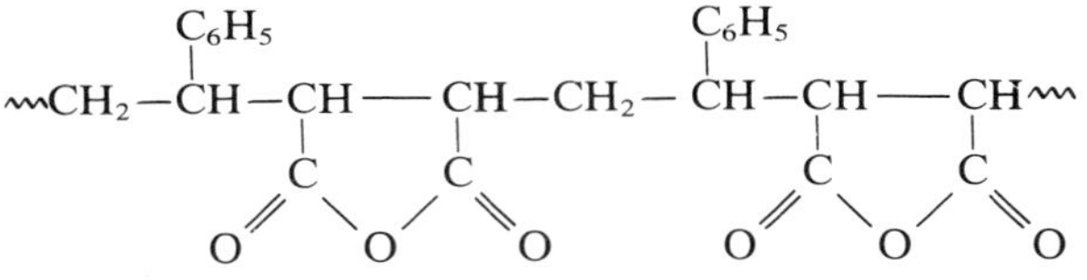

Note that it is perfectly valid, but not usual, to view this alternating copolymer as a homopolymer having the repeating unit:

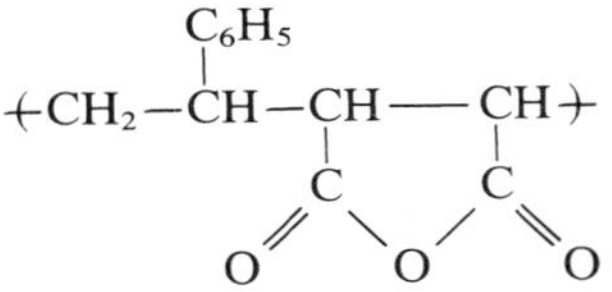

Butenedioic anhydride can also be copolymerised free radically with phenylethyne (phenylacetylene) to form an alternating copolymer having the repeating unit:

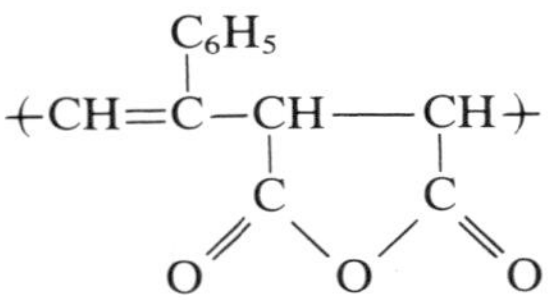

Note here that one of the monomers, phenylethyne, is in fact a triple-bonded compound (unlike the double-bonded compounds which we discussed earlier).

Block copolymers may be prepared by a variety of methods, one of which involves an anionic mechanism. In the first stage, one type of monomer is polymerised anionically and reaction is allowed to proceed until the monomer is used up. To the living polymer (see p. 14) is then added a second monomer which adds to the chain, forming the second block. The process is repeated as required. Block

copolymers of commercial interest include the phenylethene–buta-
1,3-diene block copolymers, which have the characteristics of thermoplastic rubbers (see p. 29).

Graft copolymers can be produced by free radically initiating the
polymerisation of a monomer B in the presence of the homopolymer
of a monomer A. Free radicals produced will remove atoms along
the chain of poly(A), hence producing free radical sites on the chain
itself, from which poly(B) can then grow. For example, consider the
homopolymer, $+CH_2-CHX+_n$, treated with monomer B in the
presence of a peroxide initiator. If we represent the free radicals
produced from the initiator as $R\cdot$, then:

$$\text{ⱳCH}_2-\text{CHXⱳ} + R\cdot \;\rightarrow\; \text{ⱳCH}_2-\overset{\cdot}{\text{C}}\text{Xⱳ} + RH$$

then

$$\text{ⱳCH}_2-\overset{\cdot}{\text{C}}\text{Xⱳ} + nB \;\rightarrow\; \text{ⱳCH}_2-\overset{\overset{\displaystyle (B+_n}{|}}{\text{C}}\text{Xⱳ}$$

Alternatively, irradiation with u.v. light can be used to generate the
free radicals along the homopolymer chain. Of industrial importance,
graft copolymers involving propenenitrile, buta-1,3-diene and
phenylethene have been produced.

Polymerisation techniques

Methods for preparing condensation polymers vary widely because
of the many different types of reaction which fit into this category.
However, some general comments can be made, using a linear polyester as an example. A bulk polymerisation method is widely practised
in the manufacture of condensation polymers, where reactions are
not as exothermic as in addition polymerisations. Consequently, some
of the problems which arise for addition polymerisations when using
a bulk method (see later) do not arise for polycondensations. In the
case of linear polyester preparations, the bifunctional monomers are
melted together in equimolar proportions. The increased viscosity,
as polymerisation proceeds, is offset slightly by the elevated temperatures used. The water (or other molecule) formed, as the by-
product, is distilled off to keep the position of equilibrium to the right
hand side or forwards (see p. 16). Catalysts are sometimes used to
speed up reaction. The extent of reaction can be followed by observing
the amount of water which distils off. At the required point, reaction
is stopped; the product is discharged from the reaction vessel, and
when cool may take the form of a very viscous liquid or a solid.

Addition polymerisations can be carried out using several methods.

(a) Bulk polymerisation

In bulk polymerisation, the system essentially consists of monomer (in liquid or gaseous form) and initiator. Polymerisation of liquid monomers, such as phenylethene, raises all sorts of problems. Addition polymerisations are strongly exothermic and reaction may get out of control or even become explosive. Since polymers are usually soluble in their monomers, the viscosity of the system increases and difficulties in stirring can arise. Heat dissipation is a problem; in fact, localised overheating can lead to polymer degradation. Generally speaking, although bulk polymerisation yields a relatively pure product, it is seldom used for preparing large batches of addition polymer.

(b) Solution polymerisation

Here, the monomer is dissolved in a suitable solvent prior to polymerisation being carried out. In such a polymerising system, the solvent can assist in the dissipation of (exothermic) heat of reaction. Disadvantages include the possibility of chain transfer to solvent occurring (see p. 9) with the subsequent formation of lower relative molecular mass polymer, and the fact that the solvent has to be eventually removed from the resulting polymer. This latter problem can be overcome by using a solvent which will dissolve the monomer but not the polymer; hence, the polymer is obtained directly as a slurry.

(c) Suspension polymerisation

In this process (sometimes called pearl or bead polymerisation) the monomer (containing dissolved initiator) is dispersed as droplets in water. This is done by vigorous stirring during the reaction. Polymerisation then takes place within the droplets (in other words, each isolated droplet is effectively undergoing a tiny bulk polymerisation). The droplets are kept from sticking together by the addition of small amounts of stabilisers, such as talc or poly(ethenol) (polyvinyl alcohol). Advantages of this system include the dissipation of heat of reaction into the aqueous phase, and the fact that the polymeric product is in the form of small granules which are easily handled and relatively uncontaminated.

(d) Emulsion polymerisation

Emulsion polymerisation bears some resemblance to suspension polymerisation but differs in that soap is added to stabilise the monomer droplets. The soap also forms aggregates of soap molecules or *micelles*. These micelles solubilise some of the monomer, i.e. take some of the monomer into the interior of the micelle. Initiator, which is dissolved in the aqueous phase, diffuses into the micelles, thereby initiating polymerisation. The polymer molecule grows, taking further monomer from the aqueous phase. In this way, high relative molecular mass polymer can be formed. Again, the aqueous phase absorbs the heat of reaction.

Polymer Structure and Properties **3**

Since the use to which a particular polymer can be put depends on the structure of the polymer, it is essential that this topic is looked at in some detail. We have already mentioned some of the factors which affect polymer properties (see p. 4), including relative molecular mass and crystallinity; we shall now consider these and other factors more closely.

Factors affecting crystallinity

When representing a poly(ethene) chain in the usual or conventional way, with the tetrahedral angle between the C—C bonds,

it must be appreciated that this represents an improbable, fully extended configuration. In a dilute solution of a polymer where chains are free to move, many possible chain configurations will occur. Figure 4 shows some of the irregular or coiled forms in such a solution.

In solid polymers, however, where chains are tangled, both movement and configurations of the chains will be restricted. In addition, if the chains have sufficient regularity of structure (see later), and if interchain forces (i.e. forces between chains) are sufficiently strong, chains, or parts of chains, may approach each other closely in a parallel manner resulting in crystalline regions being formed. Such regions are shown in Figure 3 (p. 5). There are various types of interchain forces. If the chains contain highly electronegative atoms, such as oxygen, then the inductive effect (i.e. the 'pull' of electrons in the

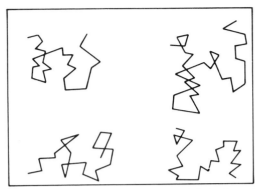

Figure 4 Possible irregular or coiled configurations of chains (or polymer molecules) in a dilute solution of a polymer.

covalent bond towards the more electronegative atom) causes the chains to become polarised. *Dipole–dipole* attractions can now occur—in other words, a $\delta-$ part of one chain can be attracted towards a $\delta+$ part of an adjacent chain. Also, a polar group on one chain may induce a dipole on an adjacent chain, again causing an interchain attraction. Attractive forces between permanent and induced dipoles are called *induction* forces, and are usually small.

A third type of force is called the *London* or *dispersion* force. If we consider an atom at a particular moment in time, electrons will not be distributed symmetrically around its nucleus—there may be more electrons on one side of the nucleus than the other. Hence, a temporary dipole exists. Similarly, such temporary dipoles exist in polymer chains and, what is more, may induce dipoles on adjacent chains resulting in interchain attractions. For polymer chains having little or no permanent dipole, attractive forces, or van der Waals forces as they are frequently called, are almost exclusively determined by London (or dispersion) forces.

A particularly important type of interchain attraction may result from hydrogen bonding, which can be considered as a specific type of dipole–dipole interaction. When a hydrogen atom is bonded to a very electronegative atom, the hydrogen nucleus is devoid of screening electrons for a large part of the time because the shared electron pair 'cloud' is distorted towards the more electronegative atom (inductive effect). The hydrogen nucleus, therefore, forms the positive end of a strong dipole ($\delta+$), and will seek out regions of high electron density such as electronegative atoms (having a $\delta-$ charge) situated on adjacent chains. This results in strong interchain attractions. An example of this occurs between the carbonyl and $\geqslant$N—H groups, on different chains, in polyamides such as nylon 6,6—see Figure 5. Further important structural effects, such as in proteins (see p. 82), also result from hydrogen bonding which, in some cases, can occur between groups on the same chain—this is called intramolecular bonding in contrast to intermolecular bonding which occurs between groups on different chains.

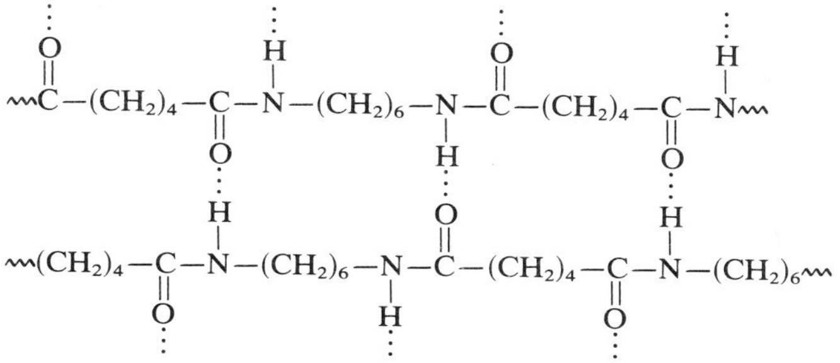

Figure 5 Hydrogen bonding between molecules of nylon 6,6 (where the dotted lines represent hydrogen bonds).

Because of the differing lengths, and tangling, of polymer molecules in a sample, very few polymers are completely crystalline. The term crystalline polymer, therefore, refers to a polymer having an appreciable degree of crystallinity, with many crystalline regions in the sample. As was mentioned earlier (p. 5), chains must be able to approach each other sufficiently closely if strong interchain forces are to operate. If this is to occur, regularity of molecular structure is an important requirement—in fact, it is perhaps the most important requirement if the polymer is to be crystalline. Consider, for example, a linear polyester (a polymer containing the $-\overset{\parallel}{\underset{O}{C}}-O-$ linkage) such as 'Terylene'. Regularity of chain structure enables chains to approach each other closely; hence, interchain forces (in this case, dipole–dipole forces) can operate, maintaining an ordered structure—see Figure 6.

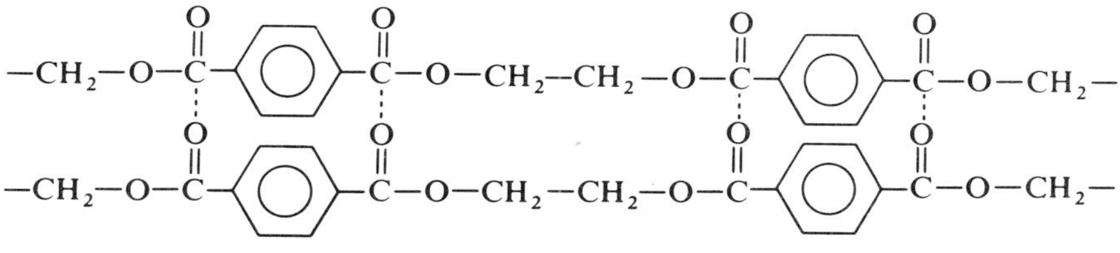

Figure 6 The regularity of chain structures in 'Terylene'. The dotted lines represent the dipole–dipole forces between carbonyl groups on adjacent chains.

It is clear that irregularities in chain structure, such as branching, will hinder chains from approaching each other closely, and crystallinity will therefore be limited.

Polymer *tacticity* is also important in determining order in a polymer sample. For example, in the polymer formed from the monomer $CH_2=CHX$, the X group can take a random configuration (*atactic* polymer), an alternating configuration (*syndiotactic* polymer) or the same configuration (*isotactic* polymer) along the chain; these different forms are shown in Figure 7. It is observed that atactic poly(phenylethene) (polystyrene), for example does not crystallise because the bulky phenyl groups (which correspond to X in Figure 7) arranged randomly along the chain prevent close approach of adjacent chains. In contrast, however, the isotactic form is capable of close approach and high crystallinity. Stereospecific polymers, such as those depicted in Figures 7(b) and (c), can often be produced using Ziegler–Natta catalysts (see p. 14).

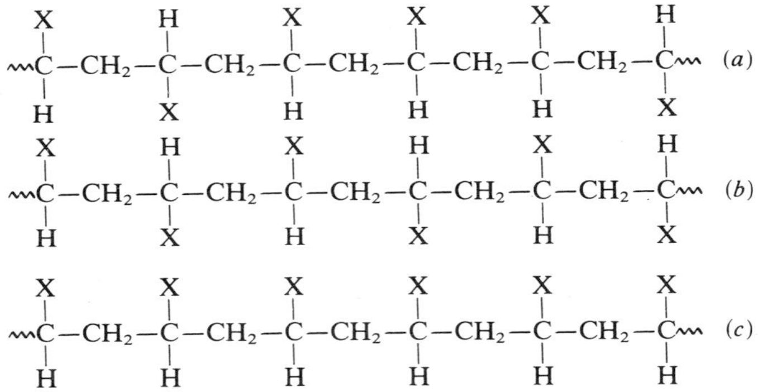

Figure 7 Representation of polymer tacticity showing (a) atactic, (b) syndiotactic and (c) isotactic polymers.

We can now consider what happens to a molten polymer when it is cooled. As the temperature is reduced (and the viscosity of the molten polymer increases), the thermal motion of the tangled polymer chains decreases until at some stage the attractive forces between chains overcome the kinetic energy of the thermal motion. If chains have regularity of structure and are mobile enough to align themselves as the temperature drops, crystallinity can be introduced into the polymer. However, even if the chains are regular in structure, some of the disorder of the molten state can be 'frozen in' as the polymer solidifies, resulting in amorphous regions being formed. When a polymer which is capable of crystallising is in the molten state, it is very often a viscous and quite transparent liquid. As it cools and crystallises it can become translucent or milky because of the presence of tiny crystallites produced in the solid polymer. Note also that because these crystallites are very small, any particular polymer chain along its length may pass through several crystallites. Because of the closer packing of chains, development of crystallinity is accompanied by an increase in density of the polymer sample.

Effects of crystallinity on polymer properties and uses

Because polymer samples can contain both crystalline and amorphous regions, it is found that the properties of a polymer can change considerably over a small temperature range as a result of two different processes associated with these two regions. Firstly, changes in polymer properties can result from the *melting* of crystallites. Melting is essentially the separation of chains in the crystalline regions, hence enabling the polymer to flow. Melting points will therefore depend on interchain forces, e.g. linear poly(ethene) (polyethylene or polythene) has a melting point (T_m) of 135 °C whereas 'Terylene', because of attractions between polar groups on adjacent chains (see Figure 6), has a melting point of 265 °C. Similarly, polyamides have high melting points because of interchain hydrogen bonding (see Figure 5). Also, an increase in chain stiffness has a tendency to raise the melting point of a polymer; hence, the introduction of an inflexible group, such as a benzene ring, into a chain can cause a marked increase in melting point.

Secondly, polymer properties can change as a result of the *glass transition*, which is associated with the amorphous regions of the polymer. Because of this transition, the polymer can change from a hard, brittle, glass-like substance to a soft, flexible, rubbery material as the temperature is raised through the *glass transition temperature* (T_g)—the effects are greatest with completely amorphous polymers. In such polymers, at temperatures below T_g, the disordered chains are 'frozen' in position and the polymer is glassy or brittle. As the temperature is raised and T_g is approached, segments of chains can now move with respect to one another; above T_g, the polymer becomes more flexible, often showing rubber-like properties since extension and contraction of chains can occur.

Any polymer, therefore, which has both crystalline and amorphous regions can have T_m and T_g values associated with it. As with T_m, increasing interchain forces and chain stiffness raise T_g since the movement of chain segments is restricted. Similarly, cross-linking makes movement of chain segments more difficult and again results in an increase in T_g (since more energy is needed for movement of chain segments to occur). In contrast, T_g can be lowered by the use of a *plasticiser*, which is a substance deliberately added to the polymer to reduce interchain forces and hence make movement of chain segments easier. In this way, poly(chloroethene) (polyvinyl chloride or PVC) can be converted from a tough, hard substance to a material which is flexible at ordinary temperatures, enabling it to be used for making articles such as raincoats, etc.

The degree of crystallinity greatly affects the properties of the polymer and therefore the uses to which the polymer can be put. For example, strength and rigidity are associated with the crystalline regions, and the development of crystallinity is important in the production of fibres. However, *orientation* is also important in fibre production. Consider a crystallisable polymer cooling from the melt; the crystallites will be oriented in every possible direction (see Figure 8(a)). If, however, the polymer is stretched, the crystalline regions become aligned or *oriented* in the direction of the applied stress (see Figure 8(b)). In this *cold-drawing* process, the sample is not gradually

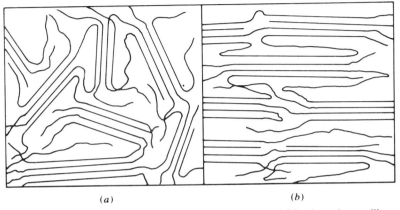

(a)   (b)

Figure 8 Schematic representation of (a) unoriented and (b) oriented crystallites.

reduced in thickness, but is found to suddenly become thinner at one point; this is known as *necking down*. The degree of crystallinity does not tend to be increased by cold-drawing if crystallinity is already well established in the sample—only the alignment of the crystallites is affected. If, however, the undrawn sample is largely amorphous, cold-drawing can increase its degree of crystallinity. Most synthetic fibres, including nylon 6,6, are subjected to cold-drawing. In this way, the tensile strength of the materials (in the direction of the fibre axis) is increased. It can now be seen why nylon and 'Terylene' lend themselves to fibre production, i.e. regularity of structure and high interchain forces encourage formation of crystallinity (see p. 25).

Crystallinity, however, is a disadvantage in the production of some materials. Consider, for example, a polymer which is to be used as an elastomer, i.e. a material having rubber-like elasticity. Here, extension and contraction is only possible if the polymer is amorphous and above its T_g; segments of chains must be able to move so that the material can extend and contract rapidly. However, chains must not slip past each other on stretching otherwise the material will not regain its original shape. Slipping of chains can be prevented by introducing cross-links into the material, a process known as *vulcanisation*. An elastomer can, therefore, be described as an amorphous high polymer, which is above its T_g, and contains cross-links to prevent slipping of chains.

Some elastomers or rubbers have no crystallinity when relaxed but develop crystallinity on stretching, hence giving strength to the rubber just when it might be expected to give way under tension. One interesting application of crystallinity in polymers is found in the production of thermoplastic rubbers (see p. 21). Here, a block of crystallisable polymer, such as poly(phenylethene) is attached to each end of a 'rubbery' polymer chain such as poly(buta-1,3-diene) (polybutadiene). When the material is at a temperature above the T_m of the crystallisable part it will be a viscous liquid, capable of being moulded by the usual techniques applied to thermoplastics, e.g. injection moulding, where polymer, softened by heat, is injected under pressure into a cool mould where it hardens to give the required article. When the polymer has cooled, however, the material will have flexibility because of the amorphous 'middle' parts of the chains, but the ends of the chains will crystallise (see Figure 9), hence acting as

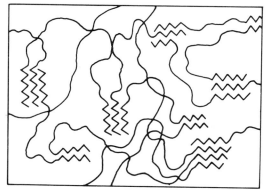

Figure 9 A thermoplastic rubber.

the occasional 'cross-links' which are needed to give the material the strength to resist creep (or slipping of chains). One disadvantage of these materials is that since the 'cross-links' are not chemical cross-links, as in the vulcanisation process just described, the thermoplastic rubber will soften completely if the temperature exceeds the T_m of the crystallisable portion. Thermoplastic rubbers are used in the manufacture of toys, elastic bands, etc.

To summarise, therefore, it can be said that polymers with symmetrical repeating units and high interchain forces can be used for fibre production, having high crystallinity and tensile strength. Plastics, however, have a lower degree of crystallinity and can be softened or moulded at higher temperatures (having T_g values above room temperature) unless they are highly cross-linked. Finally, elastomers are amorphous polymers, above their T_g at room temperature, and contain cross-links to prevent gross slipping of chains.

Determination of polymer crystallinity and structure

One of the most powerful methods for studying regular or orderly arrangements of atoms or molecules is by X-ray diffraction. X-rays are electromagnetic waves, similar to light waves but of much shorter wavelength, and are produced when a metal target is bombarded by high energy electrons. Electromagnetic radiation gives interference effects (see Figure 10) with structures comparable in size to the

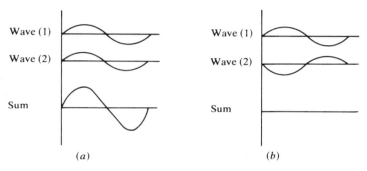

Figure 10 (a) Constructive and (b) destructive interference in wave motion.

wavelength of the radiation. If the structures are arranged in an orderly array or lattice, the radiation is reinforced only under certain experimental conditions, and knowledge of these conditions gives valuable information concerning the arrangement of the structures.

Consider a set of parallel planes of diffracting points (e.g. atoms) on which is falling a beam of X-rays of wavelength λ (see Figure 11).

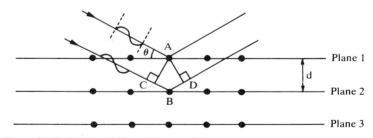

Figure 11 Reflection of X-rays by crystal planes.

If the reflected waves are still to be in phase, the extra distance travelled by the lower wave must be equal to a whole number of wavelengths, i.e. $CB + BD = n\lambda$ where n is an integer. Since $AB = d$

(the lattice spacing), then $CB = BD = d \sin \theta$. In other words:

$$CB + BD = 2d \sin \theta$$

or

$$2d \sin \theta = n\lambda \qquad (3.1)$$

Equation 3.1 is called the Bragg equation, and maximum reinforcement occurs for certain values of θ corresponding to $n = 1, 2, 3$, etc. The diffracted beam can be detected by its action on a photographic film or plate.

In an X-ray diffraction experiment, monochromatic X-rays are used. Clearly, a single crystal placed randomly in such a beam will only give strong diffraction if the orientation of the crystal with respect to the beam (i.e. the angle θ) satisfies equation 3.1. A stationary, randomly placed crystal in the beam will rarely satisfy this condition, but if the crystal is rotated, orientations corresponding to different sets of diffracting planes will be obtained resulting in strongly diffracted beams. If the latter are allowed to fall on a photographic film or plate, a pattern of spots is obtained, and measurements taken from X-ray diffraction patterns can give information concerning the dimensions of the lattice. Alternatively, a powder containing fine crystals can be used instead of a single crystal. In such a sample, some of the crystals will be oriented in such a way as to satisfy equation 3.1, and each set of reflecting planes gives rise to a cone of diffracted rays which produce a ring on a photographic plate.

With polymers, the situation can be more difficult since unoriented polymer samples, for example, contain randomly oriented crystallites together with amorphous regions (see Figure 8(a))—symmetry in polymers, therefore, is generally low compared with that in low relative molecular mass crystals, and X-ray diffraction patterns of unoriented polymers do not provide much information concerning polymer structure. However, polymers can be oriented by drawing (see p. 28), thereby aligning the crystallites in the direction of drawing (Figure 8(b)). X-ray diffraction patterns from this sort of polymer sample usually consist of discrete spots or arcs (less sharp, however, than those obtained from single crystals). Information which can be obtained from such diffraction patterns includes the determination of chain configurations in crystallites, estimations of crystallite size, and the relative proportions of crystalline and amorphous regions in the polymer sample.

Polymer degradation

A common observation made by most people is that organic materials decay—fallen leaves rot to form compost, timber deteriorates, oil in a car engine breaks down during use and impairs its lubricating powers, and so on. Polymers also deteriorate or change over a period of time, and although it is difficult to define the term degradation, perhaps a polymer technologist would find it most useful to consider degradation as any change in the properties of the material. For example, rubbers may lose their elasticity and become flaccid or even sticky on the surface, or again they may become brittle and fail under oscillating loads (such as in mountings for mechanical parts). Polymers, generally, may discolour or exhibit cracks or surface crazing when exposed to various situations (we all know the example of the 'plastic' greenhouse which starts out transparent but isn't after twelve months or so!). Clearly, there are many situations where failure of a material could have serious implications, and so a study and under-standing of degradation is important.

Various factors can produce degradation in polymers, including thermal energy, mechanical energy, radiation (u.v., etc.) and chemical agencies (oxidation, etc.). For example, some polymers, when heated, decompose by a stepwise loss of monomer units in a reaction which is essentially the reverse of polymerisation:

$$\text{monomer} \xrightleftharpoons[\text{depolymerisation}]{\text{polymerisation}} \text{polymer}$$

Poly(methyl 2-methylpropenoate) (poly(methyl methacrylate)) will do this and if heated strongly, almost complete *depolymerisation* occurs to yield monomer (this process is described in experiment **3**, Chapter 7). Decomposition of polymers, by thermal energy, is usually negligible at ordinary temperatures since the activation energy of depolymerisation is high compared with that of polymerisation. However, at some elevated temperature, the rates of polymerisation and depolymerisation can become equal. This *ceiling temperature*, as it is called, is the temperature above which polymerisation will not occur—in other words, polymer radicals, in the presence of monomer, depolymerise rather than grow. However, some polymer systems do have very low ceiling temperatures and some monomers, which were previously thought to be unpolymerisable, can now be polymerised by carrying out reaction at low temperature.

For polymers with very high theoretical ceiling temperatures, some other form of thermal breakdown can occur rather than depolymerisa-tion. For example, poly(ethene) when heated sufficiently strongly tends to cross-link, followed by random bond scission at higher temperatures. Poly(chloroethene), on heating, loses hydrogen chloride resulting in unsaturation in the chain:

$$\text{\textasciitilde\textasciitilde CH}_2-\text{CHCl}-\text{CH}_2-\text{CHCl}-\text{CH}_2-\text{CHCl}\text{\textasciitilde\textasciitilde}$$
$$\downarrow \text{heat}$$
$$\text{\textasciitilde\textasciitilde CH}_2-\text{CH}=\text{CH}-\text{CH}=\text{CH}-\text{CHCl}\text{\textasciitilde\textasciitilde} + 2\text{HCl}$$

The development of this unsaturation can be followed by observing the resulting colour changes, which deepen from yellow to orange to red and finally to black. This problem can be serious for PVC polymer (particularly if PVC is present in a fire!) and therefore stabilisers are added to it which usually consist, for example, of metallic compounds of lead and tin.

Similarly, poly(propenenitrile) (polyacrylonitrile) discolours when heated because of the sequence of reactions shown in Figure 12. This system has been of interest in research because of its semiconducting properties.

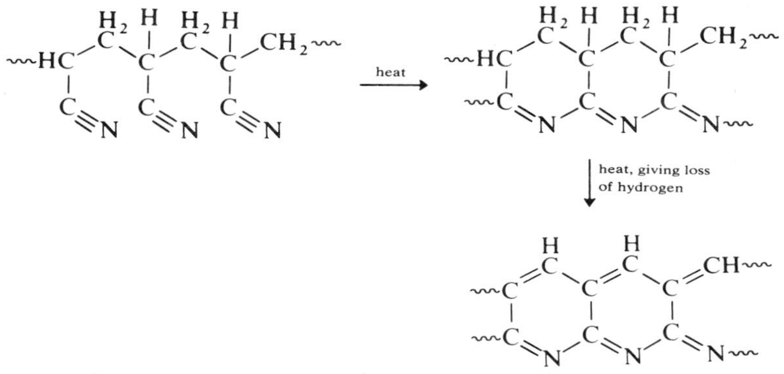

Figure 12 The thermal breakdown of poly(propenenitrile).

In a book of this size, it is impossible to give a comprehensive coverage of the topic of degradation; nevertheless, some mention should be made of the effects of sunlight and air on polymeric materials. U.v. radiation from the sun is capable of exciting electrons to higher energy orbitals, which can result in bond breakage and the formation of free radicals. Obviously, excitation will only occur if the material concerned is capable of absorbing the radiant energy and this in turn depends on the nature of the bonds and groups within the polymer. Deterioration in polymers is often caused by the joint effects of radiation from the sun together with oxygen. This can result in surface crazing and the polymer becoming brittle, probably caused by chain scission and cross-linking (the cross-linking occurs via the free radical sites produced after excitation—see above). Sometimes, transparent materials suffer from a darkening in colour because hydrogen atoms can be lost as radicals from the chain, forming gaseous hydrogen or water by oxidation, resulting in a series of double bonds being formed in the polymer rather as in the second reaction shown in Figure 12. It will be noticed that in the final structure in Figure 12, the carbon backbone of the polymer consists of alternate single and double bonds; such a structure is said to be *conjugated* and electrons are now delocalised in this conjugated structure because of orbital overlap (rather as in a benzene ring but on a larger scale). A property of more extensively conjugated structures is that electrons can relatively easily absorb radiant energy, and if absorption from white light occurs the remaining, or emerging, frequencies which reach the eye are registered as a colour. For example, we all know that hydrated copper(II) sulphate is blue—this is because d electrons in the Cu^{2+} ion absorb red light of the requisite frequencies, and white light minus red gives (peacock) blue. Substances are known which give polymeric materials protection from u.v. radiation, and it appears that they work by absorbing the damaging frequencies and re-emitting lower frequencies, which are less damaging to the polymer, or heat. It is found that poly(ethene) can be protected in this way, to some degree, by the addition of carbon black.

Interaction of polymers with liquids

The interaction of polymers with liquids is an extremely important topic, both for theoretical and technical reasons. As will be seen in Chapter 4, many polymers are characterised in solution; similarly, we have all benefited from polymers in solution when, for example, varnishing wood or painting the house. Consequently, it is not surprising to find that the research literature on this topic, and ultimately the applications of these results, is voluminous to say the least.

The factors which determine the solubility of polymers are more involved than those applicable to solubilities of low relative molecular mass materials. If common salt is added to water at a particular temperature, it will carry on dissolving until the solution becomes saturated at that temperature. For polymer–solvent systems, the picture is more complicated. Some polymers may have no saturation solubility—the polymer either dissolves completely or is just swollen by a given solvent. If the polymer does dissolve, further solubility, on adding more polymer, may be slow or even prevented by the high viscosity of the solution (see later) rather than by the attainment of saturation. Despite these complications, some general rules can be applied. Like dissolves like, so polymers and solvents having structural similarities favour solution occurring, i.e. polar polymers tend to dissolve in polar solvents. Solubilities of polymers decrease as their relative molecular masses increase. If a polymer is dissolved in a suitable solvent and a non-solvent is added (or if the polymer solution is poured into an excess of the non-solvent) the polymer can be precipitated out. This technique is used frequently in the laboratory for obtaining polymer samples from the reaction mixture, and experiment **2** in Chapter 7 illustrates this point.

If a polymer is cross-linked to any significant degree, solubility is prevented and the polymer just swells because liquid is taken into the network, hence stretching it. It is quite spectacular to watch this process the other way round. If a polymer is being produced in a solvent and cross-linking starts to occur, the viscosity of the solution increases sharply. At some point, a jelly-like material is formed which comes out of solution. This is a gel (see p. 18), and consists of a polymer network containing solvent rather like a sponge containing water. On an industrial scale, this gelation can be disastrous; for example, if too much cross-linking occurs during the production of phenol–methanal (phenol–formaldehyde) resins, where many tonnes of reactants are involved, the removal of the gel can represent a time consuming and costly mistake!

As mentioned earlier, one of the characteristics of polymers is that they produce solutions of much higher viscosity than that of the pure solvent. Viscosity measurements on polymer solutions can be used, amongst other things, for estimating relative molecular masses of polymers, and this is discussed in Chapter 4, together with the sort of equipment which is needed for these measurements. Viscosities of polymer solutions tend to decrease with decreasing concentration and increasing temperature. However, there are exceptions. Consider, for example, a polymer which contains ionisable groups, such as carboxylic acid groups. At concentrations above about 1%, chains in solution may overlap, and because of repulsion between like charges on adjacent chains, and possibly incomplete ionisation, the chains are not greatly extended. As the solution is diluted, chains become further apart—ionisation may increase as small mobile ions diffuse away, and repulsion between like charges on the same chain can now cause chains to be extended, causing an increase in viscosity. This so-called 'poly-electrolyte behaviour' can be quite startling to the chemist who has not encountered it before—the viscosity of the polymer solution decreases, as normal, as the solution is initially diluted, but as dilution is increased further, the viscosity rises again.

Other anomalous behaviour can be encountered with viscosity/ temperature measurements. Consider, for example, an aqueous solution of a base-catalysed phenol–methanal system. As the temperature is increased, the viscosity of the solution decreases as expected. At some elevated temperature, however, the viscosity starts to increase again; this is because further cross-linking now takes place (see Chapter 5, p. 67), hence resulting in an increase in viscosity.

Finally, an interesting application of viscosity theory is in the so-called 'viscostatic' motor oils. A problem with lubricating oils, in car engines for example, is that whilst the oil may have the right 'body' for lubrication whilst the engine is hot, the same oil may be too thick when the car is started in the middle of winter. The addition of suitable polymers to the oil can overcome this problem. Consider an oil which has the 'correct' viscosity when cold. A polymer which is not very soluble in this cold oil will stay coiled up, and since its effective volume is therefore small it will only have a small effect on the oil's viscosity. If, however, its solubility in the oil increases with temperature, the polymer will uncoil as the temperature of the engine, and hence the oil, rises. Therefore, the decrease in viscosity with temperature will be compensated for by the increase in viscosity due to the uncoiled polymer, and so the oil keeps its 'body' over the whole temperature range.

Relative Molecular Mass and its Determination   4

Because so many properties of a polymeric material depend on its relative molecular mass, e.g. solubility, mouldability, melt and solution viscosity, etc., a knowledge of relative molecular mass (M_r) determination is important. As we have already seen, the outcome of polymerisation reactions is that polymer molecules of various sizes are produced although, of course, two polymer molecules containing 100 repeating units of $-CH_2-CHCl-$ and 150 repeating units of $-CH_2-CHCl-$, respectively, both still constitute the polymer poly(chloroethene) (polyvinyl chloride or PVC). The main point to appreciate is that polymer samples almost invariably contain a distribution of molecular sizes and hence a distribution of relative molecular masses. Therefore, any determination of relative molecular mass will yield an average value, the two most useful averages being the number average and weight average.

The number average, $\bar{M}_n$, is defined mathematically as:

$$\bar{M}_n = \frac{\sum N_i M_i}{\sum N_i} \tag{4.1}$$

where N_i is the number of molecules with degree of polymerisation (D.P.) $= i$ and M_i is the relative molecular mass of molecules with D.P. $= i$. If relative molecular mass is determined by a colligative property, i.e. a property depending on the number of particles present, such as osmotic pressure, then a number average value, $\bar{M}_n$, is obtained.

The weight average, $\bar{M}_w$, is defined mathematically as:

$$\bar{M}_w = \frac{\sum w_i M_i}{\sum w_i} \qquad (4.2)$$

where w_i is the total mass of all molecules whose D.P. $= i$ and M_i is as above. In other words, we have averaged the relative molecular mass according to the mass of molecules of each type. Techniques such as light scattering (see below) will yield values of $\bar{M}_w$.

Let us consider a simple example to illustrate the difference between $\bar{M}_n$ and $\bar{M}_w$.

Example

Suppose a theoretical polymer sample consists of the following:

10 molecules of polymer having M_r 10 000

10 molecules of polymer having M_r 12 000

10 molecules of polymer having M_r 14 000

10 molecules of polymer having M_r 16 000

Calculate $\bar{M}_n$ and $\bar{M}_w$.

(*i*) From equation 4.1

$$\bar{M}_n = \frac{\sum N_i M_i}{\sum N_i}$$

$$= \frac{(10 \times 10\ 000) + (10 \times 12\ 000) + (10 \times 14\ 000) + (10 \times 16\ 000)}{40}$$

$$= 13\ 000.$$

(*ii*) From equation 4.2

$$\bar{M}_w = \frac{\sum w_i M_i}{\sum w_i}$$

$$= \frac{(100\ 000 \times 10\ 000) + (120\ 000 \times 12\ 000) + (140\ 000 \times 14\ 000) + (160\ 000 \times 16\ 000)}{520\ 000}$$

$$= 13\ 385.$$

As can be seen, $\bar{M}_w$ is greater than $\bar{M}_n$, and this is always true for a *polydisperse* sample, i.e. a sample which contains a distribution of relative molecular masses. If the polymer sample consists of identical molecules, then the sample is described as *monodisperse*, and under these conditions $\bar{M}_n = \bar{M}_w$ (the reader might like to verify for himself, by using equations 4.1 and 4.2, that for a monodisperse system containing 10 molecules, each having a M_r of 10 000, $\bar{M}_n = \bar{M}_w =$ 10 000). The ratio $\bar{M}_w/\bar{M}_n$ is, in fact, a measure of the polydispersity (or heterogeneity) of a polymer sample.

Fractionation of polymers

To separate a typical polymer sample, with its distribution of relative molecular masses, into groups of similar molecular mass is known as *fractionation*. Most of the methods used depend on the fact that polymer solubility decreases with increasing relative molecular mass, and we shall now look at a few of the techniques employed.

In *fractional precipitation*, the polymer sample is dissolved in some suitable solvent—solutions are usually about 0.1 per cent concentration. To this dilute solution, a non-solvent is added dropwise and with vigorous stirring. The highest relative molecular mass material becomes insoluble and separates—this is removed and further non-solvent, or precipitant, is added to precipitate polymer of the next highest relative molecular mass. This procedure is repeated until the sample has been separated into fractions of decreasing relative molecular mass. Alternatively, it is sometimes possible to find a solvent which is a good solvent for the polymer when hot but not when cold; by preparing a hot solution of the polymer and allowing it to cool, fractions of decreasing relative molecular mass will be successively precipitated.

In *fractional elution*, polymer is extracted from the solid into solution. The method used is to pack a column with, for example, glass beads which are coated with polymer. The column is eluted with solvent/non-solvent mixtures of gradually increasing solvent power. Hence, low relative molecular mass polymer emerges from the column first, followed by fractions containing gradually increasing relative molecular mass material as the proportion of solvent in the liquid mixture is increased.

Finally, a more recent technique is that of *gel permeation chromatography* (or G.P.C.). In this method, a column is prepared which contains some form of cross-linked polymeric packing material which is swollen with solvent and so contains holes or interstices. A solution of the polymer sample being investigated is then passed down the column and is eluted by the passage of more solvent. It is thought that polymer molecules find their way into the holes of the packing material in the column; the smaller polymer molecules find this the easiest and so the larger polymer molecules (having the highest relative molecular mass) are eluted from the column first since they are easily dislodged. The results from the column can be recorded by a trace on chart paper, and Figure 13 shows such a trace (which is related to the relative molecular mass distribution) for an alternating

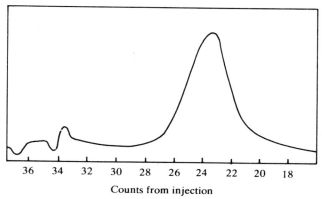

Figure 13 Gel permeation chromatogram, which is related to the relative molecular mass distribution, for an alternating copolymer formed from phenylethyne and butenedioic anhydride. (Block, Cowd and Walker (1972) Polymer, **13**, 549.)

copolymer made from phenylethyne (phenylacetylene) and butenedioic anhydride (maleic anhydride) by free radical initiation in the presence of u.v. light.

Relative molecular mass determination

Whilst we cannot hope to review all the methods for determining relative molecular masses of polymers, we shall now look at some selected examples, including some of the more important techniques such as osmometry.

(a) End-group analysis

If a particular polymer is known to contain a certain number of detectable end groups per molecule, then the number of these groups can be determined, in a known mass of polymer, by standard analytical methods. From this information, the mass of a mole of polymer can be determined and hence its relative molecular mass. To make this clearer, let us look at a simple example.

Example

Suppose 1 g of a polyester is taken which is known to contain one —COOH group per polymer molecule. If 10.00 cm^3 of standard 0.01 mol dm^{-3} sodium hydroxide solution is needed to neutralise this sample, what is the relative molecular mass of the polymer?

Moles of NaOH used are

$$\frac{10}{1000} \times 0.01 = 10^{-4} \text{ mol}$$

Therefore, number of moles of —COOH groups $= 10^{-4}$ mol. Since each polymer molecule contains one —COOH group, the number of moles of polymer present $= 10^{-4}$ mol. But 10^{-4} mol of polymer weigh 1 g. Hence, 1 mol of polymer weighs

$$\frac{1}{10^{-4}} \text{ g} = 10\ 000 \text{ g}$$

Since this method of end-group analysis has involved counting the number of —COOH groups and hence the number of polymer molecules present, the number average relative molecular mass of the polymer, $\bar{M}_n$, is 10 000.

This method has several shortcomings. Firstly, assumptions have to be made about the structure of the polymer. For example, if the polymer in the above example contained two —COOH groups per polymer molecule instead of one, or even a mixture of both types, then the calculation above would not apply. Secondly, if the relative molecular mass of the polymer is very high, then clearly 1 g of the polymer sample, for example, will contain fewer functional groups than in the example above; in fact the higher the relative molecular mass, the more difficult it becomes to determine the number of end groups present. In other words, we are limited by the sensitivity of our analytical methods. Because of this, end-group analysis has an upper working limit, for $\bar{M}_n$, of approximately 25 000.

Despite these shortcomings, the method is useful, particularly for low relative molecular mass condensation polymers. Other groups which can be easily detected analytically include hydroxyl groups (in polyesters), amine groups (in polyamides), initiator fragments in the free radically initiated polymers, and many others.

(b) Viscosity methods

The increased viscosity of polymer solutions compared with the pure solvent can be used to determine relative molecular masses of polymers. Viscosity methods have the added advantage over some other methods that experimental procedure is quick and easy, apparatus is cheap (osmometers and light scattering equipment—see pp. 43–48— are extremely expensive pieces of equipment), and the calculation of results is a simple exercise.

The usual method employed for measuring solvent and polymer solution viscosities is to use either an Ostwald or an Ubbelohde viscometer—these are shown in Figures 14(a) and (b), respectively.

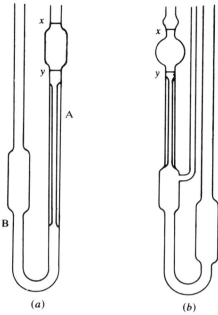

Figure 14 Capillary viscometers (a) Ostwald and (b) Ubbelohde.

Basically, we are measuring the time taken for solvent or polymer solution to flow between the two marks x and y. In the case of the Ostwald viscometer (Figure 14(a)), the volume of liquid has to be kept constant, because as liquid flows down through the capillary tube A, it has to force further liquid up limb B; consequently, if the volume of liquid is changed from experiment to experiment, the mass of liquid to be forced up limb B will change, hence yielding inconsistent flow times. In the case of the Ubbelohde, or suspended level, viscometer (Figure 14(b)), measurements are independent of the volume of liquid used since the viscometer is designed to function with liquid running just through the capillary but with no liquid underneath it (the reader will find this immediately obvious if he is fortunate enough to have access to these types of viscometers). Flow times, therefore, are measured for the solvent and for the polymer solution at various concentrations. Again, the Ubbelohde viscometer has the advantage that to achieve these various concentrations, the polymer solution can be diluted in the viscometer by adding measured amounts of solvent (since the volume of liquid used is not crucial). Measurements are taken with the viscometer clamped in a constant temperature water bath to avoid fluctations in viscosity with temperature.

Let us now look at some of the basic theory of polymer solution viscosity, and how our experimental results can be used to measure relative molecular mass. If the viscosity of the polymer solution is η and that of the pure solvent is η_0, then the *specific viscosity*, η_{sp}, of the polymer solution is defined by:

$$\eta_{sp} = \frac{\eta - \eta_0}{\eta_0} \tag{4.3}$$

and this represents the increment in viscosity caused by polymer. The ratio η_{sp}/c, where c is the concentration of the polymer solution, is called the *reduced viscosity* or *viscosity number*. The value of η_{sp}/c at infinite dilution is called the *limiting viscosity number* (this used to be known as the *intrinsic viscosity*) and is given the symbol $[\eta]$. We can express this mathematically as:

$$\lim_{c \to 0} \frac{\eta_{sp}}{c} = [\eta] \tag{4.4}$$

Since the densities of the various solutions used in any particular experiment are almost identical with the density of the solvent, it is a good approximation to assume that the viscosity of each solution produced by dilution is proportional to its flow time, and so equation 4.3 becomes:

$$\eta_{sp} = \frac{t_2 - t_1}{t_1} \tag{4.5}$$

where t_2 is the flow time for the solution whilst t_1 is that for the solvent. Experimentally, therefore, flow times for various dilutions of the polymer solution, and for the pure solvent, are obtained—values of η_{sp} and then η_{sp}/c are calculated, and these η_{sp}/c values are then plotted against their corresponding concentrations, c. A plot such as that shown in Figure 15 should be obtained, and extrapolation to zero concentration will then yield a value for $[\eta]$.

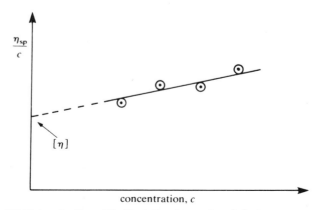

concentration, c

Figure 15 Determination of limiting viscosity number, $[\eta]$.

Mark and Houwink found that the limiting viscosity number is linked to relative molecular mass by the expression:

$$[\eta] = KM^a \tag{4.6}$$

where M is the relative molecular mass and K and a are constants for a given polymer–solvent system. Clearly, K and a have to be determined using at least two other samples of the polymer having different relative molecular masses, and these relative molecular mass values have to be measured using some absolute method such as osmometry or light scattering. If this method yields number average relative molecular masses, then the K and a values obtained will also yield number average values with any viscosity data to which they are applied, and similarly with other averages. It can be seen, therefore, that viscometry is not an absolute method for determining relative molecular mass since without knowing the values for K and a, relative molecular masses cannot be calculated. However, if this data is known, the ease of experimentation makes viscometry a very attractive method.

(c) Osmometry

Dilute solutions of polymers only give very small elevations of boiling point and depressions of freezing point, making it difficult to measure them accurately. Osmotic pressures of such solutions, however, are much easier to measure and provide a very good, absolute method for determining relative molecular masses of polymers. Since osmotic pressure is a colligative property, i.e. it depends on the number of solute particles present, osmometry therefore yields number average relative molecular mass values.

Osmosis may be defined as the passage of solvent but not solute through a *semipermeable membrane* from solvent to solution or from a solution of low to one of higher concentration. Semipermeable membranes, which allow the passage of solvent but not solute, can include materials such as parchment, cellophane and other cellulosic materials. We can demonstrate osmosis very easily using the apparatus shown in Figure 16. Initially, the levels in the solvent and solution columns are the same. On standing, however, osmosis occurs as solvent passes through the semipermeable membrane into the solution. Hence, the level in the solution column rises, but eventually stops; this is because a position of dynamic equilibrium has been reached where the resulting head of solution, h, prevents further transfer of excess solvent from solvent to solution. Movement of solvent across the membrane is still occurring in both directions, of course, but is now at the same rate.

There is sometimes some confusion as to what is meant by *osmotic pressure*. The osmotic pressure, Π, of a solution is the external pressure

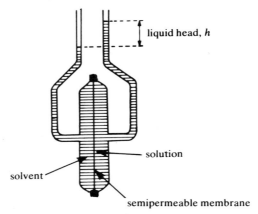

liquid head, h

solution

solvent

semipermeable membrane

Figure 16 A simple osmometer.

which must be applied to it in order to prevent passage into it of excess solvent through a semipermeable membrane. Therefore, in Figure 16, the pressure due to the head of solution, h, is not quite equal to the osmotic pressure of the original solution since osmosis, and hence dilution of the original solution, has already occurred. A more accurate measurement of Π was carried out by Berkeley and Hartley, who were amongst the pioneers in the work on osmosis; their method of measuring Π for a given solution actually prevented excess solvent flowing, and modern osmometers used in polymer characterisation laboratories are based on their original work (see later).

It was the German botanist, Pfeffer, who first made measurements of osmotic pressure on, amongst other things, solutions of sucrose. However, it was van't Hoff who used these results to deduce the laws of osmosis, namely that Π is proportional to concentration, c, and also to the absolute temperature, T:

$$\Pi \, \alpha \, c \quad \text{and} \quad \Pi \, \alpha \, T$$

Hence, on combining these equations:

$$\Pi = kcT \tag{4.7}$$

where k is a constant of proportionality. Since concentration is defined as:

$$c = \frac{\text{number of moles of solute } (n)}{\text{volume of solution } (V)}$$

then on substituting into equation 4.7:

$$\Pi = k \frac{n}{V} T$$

or

$$\Pi V = nkT \tag{4.8}$$

Further experimentation by van't Hoff showed that the constant k (for 1 mol of solute) had almost the same value as the gas constant per mole, R. Hence:

$$\Pi V = nRT \tag{4.9}$$

(cf. the ideal gas equation $pV = nRT$). Since the number of moles of solute $n = w/M$, where w is the mass of solute and M is its relative molecular mass, then on substituting into equation 4.9:

$$\Pi V = \frac{w}{M} RT \tag{4.10}$$

and since w/V = concentration of solution, c', measured in mass per unit volume, then equation 4.10 becomes:

$$\frac{\Pi}{c'} = \frac{RT}{M} \tag{4.11}$$

This is the van't Hoff equation (it can be derived from thermodynamic principles but this is beyond the scope of this text). The equation holds for ideal solutions and reasonably well for dilute solutions of small molecules, but for solutions of high polymers the equation takes the form of a power series:

$$\frac{\Pi}{c'} = \frac{RT}{M} + Bc' + Cc'^2 + \cdots \tag{4.12}$$

where B and C are called virial coefficients and are correction terms to the ideal equation given in equation 4.11. For dilute solutions where concentration is low, we can therefore neglect higher terms of c', and so equation 4.12 approximates to:

$$\frac{\Pi}{c'} = \frac{RT}{M} + Bc' \tag{4.13}$$

If the osmotic pressure is measured, therefore, for a polymer solution at constant temperature and various concentrations, a plot of Π/c' against c' will give a straight line graph of slope B and intercept RT/M (see Figure 17). Hence, the (number average) relative molecular mass, $\bar{M}_n$, of the polymer, can be calculated. It should be stressed at this point, however, that the arguments presented here have been simplified. Osmometry is a very involved topic and is very much a specialised area of research work in itself; results are very often nowhere near as simple as the above paragraphs may suggest.

Finally, we should mention the osmometer itself. The simple osmometer shown in Figure 16 is referred to as a *static* osmometer.

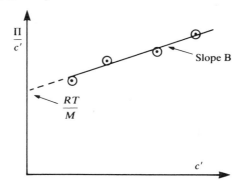

Figure 17 Estimation of the number average relative molecular mass by plotting Π/c' against c'.

Modern osmometers used widely in polymer characterisation laboratories, however, are known as *dynamic* osmometers since an automatic mechanism applies a back pressure to the polymer solution which then opposes the osmotic pressure. In other words, when the applied pressure is in equilibrium with the osmotic pressure, the applied pressure will represent Π. Hence, the dynamic type of osmometer actually prevents passage of excess solvent occurring. Despite the difficulty of selecting suitable membranes (and the cost of good quality osmometers!), osmometry remains one of the most frequently used methods for determination of the number average relative molecular mass.

(d) Light scattering

When a light beam is passed through a non-absorbing liquid, some light is scattered; this is a result of non-uniformities in the density of the liquid. If the liquid or solvent is made more inhomogeneous by the addition of solute molecules, additional scattering will occur, and this increase in scattering can be related to the concentration and the relative molecular mass of the solute. It was in 1944 that Debye extended these relationships to the determination of the weight average relative molecular masses of polymers. The theory of light scattering is extremely complex, so we will only briefly discuss the main points behind this technique. If the reader wishes to pursue this subject further, some suggested further reading is given at the end of Chapter 7.

If the polymer molecules are small in comparison to the wavelength of the light used, the following relationship is observed (which is known as the Debye equation):

$$\frac{Hc}{\tau} = \frac{1}{\bar{M}_w} + 2A_2c + \cdots \qquad (4.14)$$

where $\bar{M}_w$ is the weight average relative molecular mass, A_2 is the second virial coefficient (see equation 4.12), τ is the turbidity, defined as the fraction of light scattered in all directions from the incident primary beam per cm of path in the solution, c is the concentration, and H is given by:

$$H = \frac{32\Pi^3 n_0^2 (dn/dc)^2}{3N_A\lambda^4} \tag{4.15}$$

where n_0 is the refractive index of solvent, dn/dc is the refractive index gradient, N_A is the Avogadro constant and λ is the wavelength of the light. It follows, therefore, that a plot of Hc/τ against c for dilute solutions will be linear with a slope of $2A_2$ and an intercept of $1/\bar{M}_w$—hence, $\bar{M}_w$ is obtained for the polymer. Note that equation 4.15 is only valid for vertically polarised light (if unpolarised light is used, τ has to be multiplied by $(1 + \cos^2\theta)$ where θ is the angle of observation).

If, however, the polymer molecules are not small in comparison to λ, light scattered from different parts of the molecule will be capable of interference. Hence, for larger polymer molecules, equation 4.14 takes the form:

$$\frac{Hc}{\tau} = \frac{1}{\bar{M}_w P(\theta)} + 2A_2c + \cdots \tag{4.16}$$

where $P(\theta)$ is derived from interference theory and expresses the intensity of the scattered light as a function of the angle θ. $\bar{M}_w$ can now be estimated by doing what is known as a Zimm plot, but we will not pursue this matter further.

The basic essentials of light scattering apparatus are shown in Figure 18. Light from a mercury vapour lamp A passes through a mono-chromatising filter B and then enters the glass cell C which contains the polymer solution. It is essential to take the utmost care with the

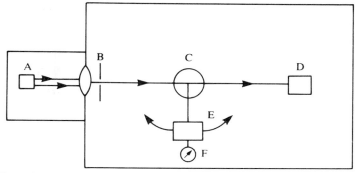

Figure 18 Basic essentials of light scattering apparatus.

solution so that it is free of dust or other scattering material—to do this, the solution is usually centrifuged or filtered through ultra-fine sintered glass filters or modern membrane filters. Transmitted light is absorbed in a light trap *D*. The intensity of the scattered light is measured by allowing it to fall on to a photomultiplier, *E*, which is mounted on a movable arm so that measurements of scattered light can be made at a variety of angles to the incident beam. The output from the photomultiplier is then measured by a galvanometer, *F*. The only other measurement which now has to be made is that of the refractive index gradient (or the rate of change of refractive index with concentration of solute), *dn/dc*. The difference between the refractive indices of solvent and solution is usually very small, and so is measured using a differential refractometer, ordinary refractometers not being sensitive enough for this. Light scattering is a very useful and flexible technique and can be used over a wide range of relative molecular masses (even in excess of 1 million). Its major drawbacks are the expense of the equipment and the complexity of the technique as a whole.

Synthetic High Polymers   5

As was mentioned in Chapter 1, the production and use of synthetic high polymers plays a major part in the economy of any modern industrialised society, and Table 2 (p. 6) gives a few examples of the output of these materials in various countries. Such is the impact that 'plastics' have had on everyday life, from window frames to heart valves, that it would be difficult to imagine how we would manage without them, and research and development is continually being carried out by the chemical industry, Government research institutions and the universities to find further applications for polymeric materials.

A useful way of classifying synthetic polymers is to split them into addition polymers and condensation polymers, and so we will also make use of this division.

Synthetic addition polymers

In the production of any polymer, it is essential that the monomer from which the polymer is to be produced should be cheap, easily obtained and easily purified if the whole process is to be a viable proposition. Hence, it is not surprising to find that many synthetic addition polymers are based on ethene (ethylene) and its derivatives, since ethene and the higher alkenes are relatively easily obtained from the cracking of petroleum:

$$C_{10}H_{22} \xrightarrow{\text{crack}} C_8H_{18} + C_2H_4$$

decane   octane   ethene

It will be convenient at this stage to classify some of the main structural units in addition polymers, using the well known, and industrial, terminology. Using the compound $CH_2\!:\!CRR'$ as our general formula, we can say:

(a) When one of the groups, R or R', is a hydrogen atom, then the monomer contains the $CH_2\!:\!CH-$ group or vinyl group, and the resulting polymer may be referred to as a *vinyl* polymer.

(b) When neither R nor R' is a hydrogen atom (e.g. as in $CH_2\!:\!C(CH_3)_2$), then the monomer contains the vinylidene group, $CH_2\!:\!C{<}$ and the resulting polymer may be referred to as a *polyvinylidene*.

Finally, polymers obtained from propenoic acid (acrylic acid), $CH_2\!:\!CH.CO_2H$, or its derivatives, are termed *acrylic* polymers.

Poly(ethene) (polyethylene or polythene)

The polymerisation of ethene to give a high relative molecular mass polymer was first accomplished by Fawcett and Gibson of Imperial Chemical Industries Ltd (UK) in 1933 using high pressure techniques. The first commercial high pressure poly(ethene) plant did not begin production, however, until 1939. High pressure processes operate at temperatures of up to about 250 °C and pressures of between 1000 and 3000 atmospheres. The mechanism is one of free radical polymerisation; peroxides, azo compounds and traces of oxygen can be used to initiate the process. Impurities, such as hydrogen, can act as transfer agents (see p. 9) and must therefore be excluded if high relative molecular mass polymer is to be obtained. As mentioned in Chapter 2 (p. 15), high pressure techniques produce poly(ethene) which is branched. However, in 1953, Professor Karl Ziegler working at the Max Planck Institute reported the polymerisation of ethene at pressures little above atmospheric pressure, using organo-metallic catalysts in conjunction with halides of transition metals. Poly(ethene) produced by this method gave a linear polymer having a higher degree of polymerisation than that produced by the high pressure process. As one might expect, the absence of branches also resulted in the low pressure poly(ethene) having a more closely packed, dense structure, with a greater rigidity and higher softening temperature. At approximately the same time as Ziegler, low pressure catalysts were also produced by the Phillips Petroleum Co. (USA) and the Standard Oil Co. (USA). The first plant using the Ziegler process began production in 1955, but the bulk of commercial poly(ethene), however, is still produced by high pressure methods.

Poly(ethene) is chemically very inert. It does not dissolve in any solvent at room temperature but is swollen by some hydrocarbons and tetrachloromethane (carbon tetrachloride). It has good resistance to acids and alkalis but can be attacked by concentrated nitric acid. Poly(ethene) ages on exposure to light and oxygen. Degradation by u.v. light can be reduced, however, by the addition of small amounts of carbon black (see p. 34). When heated sufficiently strongly, poly(ethene) cross-links followed by random bond scission at higher temperatures, but depolymerisation does not occur. Solutions or suspensions of poly(ethene) in tetrachloromethane at about 60 °C can be chlorinated to produce a soft, rubbery product, the random introduction of chlorine atoms into the chains destroying crystallinity. A more useful product called *hypalon* can be formed by chlorosulphonation, which converts the thermoplastic poly(ethene) to a vulcanisable elastomer containing about 30 per cent chlorine and 1.5 per cent sulphur. Vulcanisation is generally carried out by heating with certain metal oxides; the final product has good resistance to chemical attack, weathering and cracking.

The physical properties of high and low pressure poly(ethenes) are very different. As already stated, branching results in a lowering of crystallinity, density, softening point and crystalline melting point. The low density poly(ethene), therefore, with its flexibility, high tear strength and resistance to moisture and chemicals generally, lends itself to a wide range of uses such as for film or foil packaging, containers, electrical insulation, as sheathing for wires and cables, and many others. The high density poly(ethene) with its greater tensile strength and stiffness is used for piping, tubing and a whole range of objects and vessels.

Other polyalkenes (polyolefins)

It is not surprising that in view of the success of poly(ethene), it was not long before attempts were made to polymerise propene (propylene), $CH_2=CH-CH_3$. In Chapter 2, we mentioned Natta's work on the polymerisation of propene. Working at Milan Polytechnic, Professor Natta found that by using Ziegler-type catalysts, stereospecific polymers of propene could be produced. In particular, he synthesised isotactic poly(propene) (polypropylene) (if the reader refers to Figure 7 (p. 26) and allows the X group to be a methyl group, then Figure 7(c) will represent isotactic poly(propene)).

Because of the stereoregularity of this polymer, chains can pack together more closely, resulting in a strong, heat-resistant plastic. Properties such as tensile strength and stiffness, whilst similar to high density poly(ethene) at room temperature, are maintained to a greater degree at higher temperatures. Solubility characteristics of poly(propene) are similar to those of poly(ethene), being insoluble at room temperature.

Poly(propene) products are more resistant to scratching than the corresponding products of poly(ethene). Poly(propene) has found uses in such products as washing-machine parts, car components, chairs, handles, crates and boxes, pipes, electrical insulants, and packaging (in the form of films) for foods and goods.

Polymers derived from 2-methylpropene (isobutylene) are also of importance industrially. The monomer can be polymerised with Friedel–Crafts catalysts (see p. 11) to give, provided that the relative molecular masses are high enough, rubbery solids. They are used to some extent in adhesives and sealing compounds, but the largest use of 2-methylpropene is in the production of butyl rubber which is an elastomer made by copolymerising 2-methylpropene with methyl-buta-1,3-diene (isoprene).

Polydienes

Various structures can be produced from the polymerisation of 1,3-dienes, such as buta-1,3-diene (butadiene), depending on whether addition occurs in the 1,2- position:

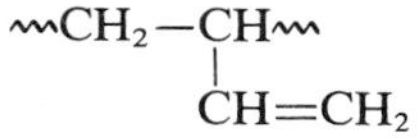

$$\sim\!\!\sim\!CH_2-\underset{\underset{CH=CH_2}{|}}{CH}\!\sim\!\!\sim$$

or in the 1,4- position:

$$\sim\!\!\sim\!CH_2-CH=CH-CH_2\!\sim\!\!\sim$$

Alternatively, both types of addition may occur together in a random manner. Note that the 1,4- addition results in the formation of *cis*- and *trans*- isomers (Figure 19).

Figure 19 Formation of (a) *cis*- and (b) *trans*- isomers resulting from (1,4-addition) polymerisation of buta-1,3-diene.

As has already been mentioned, buta-1,3-diene was originally polymerised by an anionic process using metallic sodium as catalyst (see p. 13)—the products were called Buna rubbers (deriving their name from the first letters of butadiene, as it was called then, and the symbol Na for sodium). However, improvements to these rubbers were later made, most notably by copolymerising buta-1,3-diene with phenylethene (styrene) to yield an elastomeric copolymer—these rubbers were produced in large quantities during World War II under the name of Buna-S (for obvious reasons). The buta-1,3-diene–phenylethene rubbers used now are known as SBR polymers and are prepared by emulsion polymerisation. The tensile strength of SBR polymers is similar to that of natural rubber, and they tend to have a greater resistance to oxygen and trioxygen (ozone) attack. They are used in the manufacture of tyres (see also block copolymers of buta-1,3-diene–phenylethene, p. 29).

Buta-1,3-diene can also be copolymerised with propenenitrile (acrylonitrile), again using an emulsion system. The resulting elastomers, which can be vulcanised with sulphur, are referred to as nitrile rubbers, and have excellent resistance to heat, oils and solvents. Hence, they can be used for hose, tank linings and gaskets.

Elastomeric polymers and copolymers of 2-chlorobuta-1,3-diene (chloroprene) are referred to as neoprenes. They can be prepared by emulsion polymerisation in the presence of sulphur—vulcanisation can be carried out by heating with zinc and magnesium oxides. Neoprenes have a high degree of oil and solvent resistance, but not as good as nitrile rubbers. They have a good resistance to most chemicals, oxygen and trioxygen, however. Neoprenes find applications in hose, gaskets and weather strips.

Finally, it should be mentioned that processes have been developed which have enabled methylbuta-1,3-diene (see Figure 20(a)) to be polymerised yielding a polymer almost indistinguishable from natural rubber. As discussed earlier, buta-1,3-diene can undergo addition in the 1,2- or the 1,4- position; not only that, the 1,4- addition product exhibits *cis*- and *trans*- isomers and, similarly, methylbuta-1,3-diene can also do this. Natural rubber is a high polymer of methylbuta-1,3-diene but, as is now clear, this polymer will be capable of exhibiting isomerism (see Chapter 6, p. 87). Natural rubber has the *cis*-1,4 structure (see Figure 20(b)), and methylbuta-1,3-diene can now be polymerised to give this isomer using a dispersion of lithium in the monomer as catalyst—other methods employ Ziegler catalysts.

Figure 20 The structures of (a) methylbuta-1,3-diene and (b) the *cis*-1,4 isomer of poly(methylbuta-1,3-diene) (polyisoprene).

Poly(phenylethene) *(polystyrene)*

Most phenylethene monomer is prepared by the following route:

$$\text{benzene} \xrightarrow[\text{AlCl}_3,\ 95\,°C]{C_2H_4} \text{ethylbenzene (}CH_2\text{—}CH_3\text{)} \xrightarrow[\text{steam, } 630\,°C]{-H_2} \text{phenylethene (}CH\text{=}CH_2\text{)}$$

The difficulty is in purifying the phenylethene by distillation, since the monomer is susceptible to polymerisation at even moderate temperatures. Therefore, special columns are used, an inhibitor (commonly sulphur) is added, and distillation is carried out under reduced pressure.

Phenylethene can be polymerised using heat, light or catalysts. The degree of polymerisation of the polymer depends on the polymerisation conditions—very high polymers can be produced by using temperatures little above room temperature. As has already been mentioned (p. 26), isotactic poly(phenylethene) can be produced using Ziegler catalysts, and these polymers can form highly crystalline materials.

Poly(phenylethene) is a clear, transparent (unless fillers or pigments have been added) thermoplastic material which softens at about 100 °C. It is resistant to acids, alkalis and other corrosive materials, but is readily soluble in aromatic and chlorinated hydrocarbons and is swollen, but not dissolved, in propanone (acetone). It is affected by long exposure to u.v. light or heat, and has a rather low impact strength and heat resistance. It is widely used for mouldings, coatings and sheets. Expanded or foamed poly(phenylethene) is obtained by heating poly(phenylethene), containing a gas-producing or blowing agent, with steam to give a low density foam; a system often employed is that of poly(phenylethene) beads impregnated with a volatile hydrocarbon. When the beads are heated in steam, they soften and the volatilisation of the hydrocarbon together with the diffusion of steam into the beads causes the beads to expand. Expanded poly(phenylethene) is used in large quantities as a thermal insulation material and for packaging delicate articles.

The use of phenylethene in copolymers has already been discussed.

Poly(chloroethene) *(polyvinyl chloride or PVC)*

Poly(chloroethene) is the third of the 'big three' thermoplastics, the other two being poly(ethene) and poly(phenylethene). The monomer can be prepared by two main methods.

(*a*) From ethyne (acetylene), using the catalytic addition of HCl:

$$CH{\equiv}CH + HCl \xrightarrow[\text{HgCl}_2]{180\,°C,\ 5\ atm} CH_2{=}CHCl$$

(*b*) From ethene:

$$C_2H_4 \xrightarrow{Cl_2} CH_2Cl{-}CH_2Cl \xrightarrow[\sim500\,°C]{\text{pyrolysis}} CH_2{=}CHCl + HCl$$

1,2-dichloroethane
(ethylene dichloride)

The polymer is usually prepared by suspension or emulsion polymerisation at temperatures between 20 °C and 50 °C. Polymers have good resistance to acids and alkalis, and are non-inflammable. They are relatively unstable to heat and light, and darkening of colour due to evolution of HCl does take place (see p. 32). The pure polymer is virtually unknown, therefore, outside the vinyl industry since it is normally treated with a variety of stabilisers, plasticisers (see p. 27), pigments, and so on, to give a whole range of polymers with differing physical and chemical properties. Uses of treated PVC polymers include the manufacture of pipes, sheeting, tubing, rainwear, cable insulation, bottles, gramophone records, toys, flooring and many others.

Copolymers of chloroethene (vinyl chloride) are also produced in large quantities, the more important ones involving propenenitrile and ethenyl ethanoate (vinyl acetate) as comonomers (see p. 19).

Poly(ethenyl ethanoate) (polyvinyl acetate or PVA)

Ethenyl ethanoate is a colourless liquid, b.p. 72 °C, and is produced by reacting ethyne with ethanoic acid (acetic acid). It can be polymerised in solution or, more usually, by emulsion polymerisation. Low relative molecular mass polymer tends to be soft and gummy, but polymers having a higher degree of polymerisation are harder and less flexible. The polymer is stable to normal temperatures and to light, and is soluble in aromatic solvents and polar organic liquids. The most important application for poly(ethenyl ethanoate) is in the adhesives industry, although a large amount is used in water-based 'emulsion' paints. Adhesive films and coatings of PVA are tough and wear-resistant—with a filler, PVA can be used for wall tiles. A copolymer of ethenyl ethanoate and butenedioic anhydride (maleic anhydride) has been used as a soil conditioner.

Poly(ethenol) (polyvinyl alcohol)

Although the monomer of poly(ethenol) cannot be isolated, the polymerised form is stable and is produced from PVA (see Figure

21). The reaction is one of alcoholysis and is acid- or base-catalysed, the alkaline alcoholysis being more rapid. One method involves swelling the PVA with methanol in the presence of potassium hydroxide or mineral acid; on heating, the poly(ethenol) is precipitated:

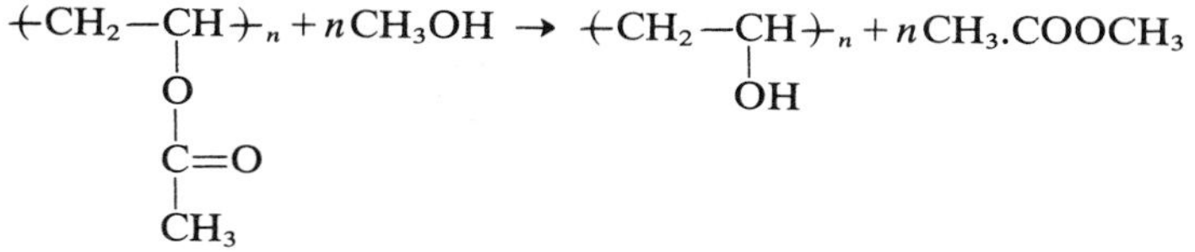

$$\begin{array}{c} \text{+CH}_2-\underset{\displaystyle \underset{\displaystyle \underset{\displaystyle \text{CH}_3}{C=O}}{O}}{\text{CH}}\text{+}_n + n\,CH_3OH \end{array} \rightarrow \begin{array}{c} \text{+CH}_2-\underset{\displaystyle OH}{CH}\text{+}_n + n\,CH_3.COOCH_3 \end{array}$$

Figure 21 Production of poly(ethenol) by alcoholysis.

Because the hydroxyl groups are small, they can easily fit into the polymer lattice resulting in crystalline polymers similar in configuration to poly(ethene). As one might expect from the structure of poly(ethenol), hydrogen bonding occurs between hydroxyl groups on different chains. Although the polymer is water-soluble (because of the hydroxyl groups), heat may be required to break the interchain hydrogen bonds to obtain solution.

The toughness of poly(ethenol) films, their resistance to chemical attack and impermeability to gases makes them useful as protective coatings. Plasticised with such liquids as propane-1,2,3-triol (glycerol), poly(ethenol) can be used to make tubing.

Poly(methyl 2-methylpropenoate) (poly(methyl methacrylate) or 'Perspex')

Methyl 2-methylpropenoate (methyl methacrylate) is prepared from propanone by the route shown in Figure 22. As can be seen, this monomer is the methyl ester of 2-methylpropenoic acid (methacrylic acid), $CH_2{=}C(CH_3)-COOH$.

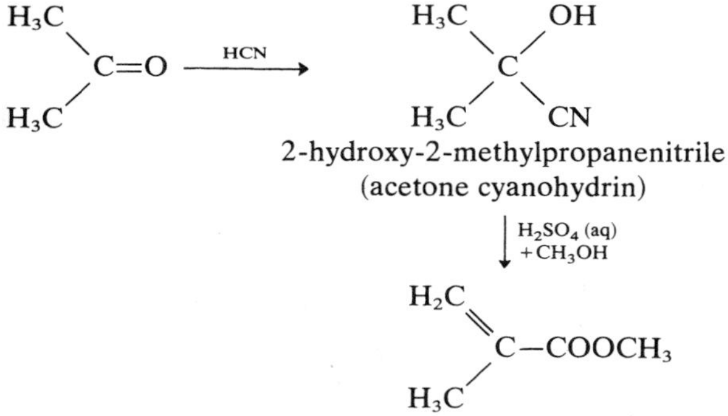

$$\begin{array}{ccc} H_3C & & H_3C \quad OH \\ \diagdown & \xrightarrow{\text{HCN}} & \diagdown \diagup \\ C{=}O & & C \\ \diagup & & \diagup \diagdown \\ H_3C & & H_3C \quad CN \end{array}$$

2-hydroxy-2-methylpropanenitrile
(acetone cyanohydrin)

$$\Big\downarrow \begin{array}{l} H_2SO_4\,(aq) \\ + CH_3OH \end{array}$$

$$\begin{array}{c} H_2C \\ \diagdown\diagdown \\ C-COOCH_3 \\ \diagup \\ H_3C \end{array}$$

Figure 22 Preparation of methyl 2-methylpropenoate from propanone.

Polymerisation of the monomer occurs easily under the influence of heat, light or initiators (the reader can prepare this polymer by trying experiment **1** in Chapter 7); an inhibitor, often benzene-1,4-diol (hydroquinone), is added to prevent polymerisation during storage or transport of the monomer. Polymerisations in industry are carried out using peroxide free radical initiators; suspension polymerisation is used for polymers which are to be moulded, but sheets or rods are often obtained by bulk polymerisation in suitably shaped vessels (casting). Since shrinkage occurs during polymerisation, casting is carried out using a 'syrup' of partially polymerised monomer; in this way, shrinkage is reduced and less heat is evolved during polymerisation.

Poly(methyl 2-methylpropenoate) is a hard, rigid transparent substance. An outstanding property of 'Perspex' is its clarity; it absorbs very little visible light, making it very useful in a wide range of applications needing high light transmission, such as windshields, etc. It has good mechanical and thermal properties—it can be drilled, machined, sawn and cemented, but it has the disadvantage of being easily scratched. Its softening temperature of just over 100 °C means that it is an easily moulded thermoplastic. Other uses include light fittings, display signs and for producing dentures.

If heated sufficiently strongly, 'Perspex' almost completely depolymerises to yield monomer (see experiment **3**, Chapter 7), making the recovery of monomer from scrap polymer into a feasible proposition.

Poly(propenenitrile) (polyacrylonitrile)

Propenenitrile can be obtained by passing hydrogen cyanide and ethyne over a catalyst of sodium cyanide at approximately 500 °C*.

$$CH{\equiv}CH + HCN \xrightarrow[{500\,°C}]{NaCN} CH_2{=}CH{-}CN$$

Polymerisation is best carried out by suspension or emulsion techniques using an aqueous redox system; the reaction is catalysed by traces of metal ions, such as Cu^{2+} and Fe^{3+}. The polymer, like 'Perspex', is classed as an acrylic polymer. It is hard, and not easily moulded or plasticised. Interchain forces are strong, yielding crystalline polymers with a melting point of about 300 °C. Poly(propenenitrile) tends to discolour on heating, following the scheme shown in Figure 12 (p. 33).

---

* An alternative process, known as the Sohio process, uses the oxidation of a mixture of propene and ammonia with air, and is now the preferred method since it is based on a cheap hydrocarbon.

The polymer is used mainly in the production of fibres, such as 'Orlon' and 'Acrilan'. These are fibres of poly(propenenitrile) containing a minor proportion (approximately 10 per cent) of one or two comonomers, such as ethenyl ethanoate. The fibres are relatively insensitive to moisture and have good weather resistance. Garments made from these fibres will dry quickly, if wet, without shrinking or stretching.

Poly(tetrafluoroethene) (*PTFE or 'Teflon'*)

The monomer is prepared by firstly producing chlorodifluoromethane, followed by pyrolysis:

$$CHCl_3 \quad + HF \xrightarrow[65\,°C]{SbCl_5} \left. \begin{array}{l} CHFCl_2 \\ CHF_2Cl \\ CF_2Cl_2 \end{array} \right\} + HCl$$

trichloromethane
(chloroform)

(mixture of chlorofluoromethanes)

then

$$2CHF_2Cl \xrightarrow{650–800\,°C} CF_2{=}CF_2 \qquad + 2HCl$$

chlorodifluoromethane   tetrafluoroethene

The monomer, which is a gas liquefying at $-76\,°C$, may be polymerised by a suspension method using fairly high pressure and a peroxide initiator. PTFE is a white solid with a waxy feel and appearance. It is a tough, flexible material having very good resistance to chemical attack. It is non-inflammable and is a first class electrical insulator. It does not weather, but is degraded by high energy radiation.

Its most useful properties are its electrical insulation, and its self-lubricating and non-stick characteristics. It is widely used for gaskets, seals, protective linings, non-lubricated bearings, 'non-stick' surfaces to prevent adhesion of food, glues, etc., and of course in electrical insulation for wires and cables. The reader may well have noticed 'Teflon' sleeves to prevent locking of stoppers and joints, such as in burettes and also round moving stirring rods, in chemical apparatus.

Synthetic condensation polymers

We have already looked at the basis of condensation polymerisation in Chapter 2; we will now consider a few examples of industrially important materials and their preparations and uses. Typical condensation polymers include the polyesters, polyamides and polyurethanes—these polymers have characteristic linkages and are shown in Figure 23.

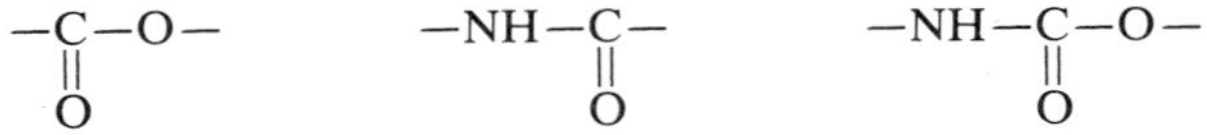

$$-\underset{\underset{O}{\|}}{C}-O- \qquad -NH-\underset{\underset{O}{\|}}{C}- \qquad -NH-\underset{\underset{O}{\|}}{C}-O-$$

(a)   (b)   (c)

Figure 23 Characteristic linkages in (a) polyesters, (b) polyamides and (c) polyurethanes.

To remind ourselves again, the special requirement for polymerisation to occur is that the monomers used must be polyfunctional (see p. 3). If each monomer contains only two functional groups, linear polymers are produced. If, however, the functionality of one or more of the monomers exceeds two, then network structures can be formed (p. 17).

It is also worth noting that relative molecular masses of synthetic condensation polymers are usually less than values obtained for addition polymers, often lying within the range 10 000 to 30 000.

Polyesters

Linear polyesters can be produced by heating together suitable dibasic acids and dihydric alcohols, as illustrated by the following reaction:

$$HO-R'-OH + HOOC-R-COOH \rightarrow$$

$$HO-R'-OOC-R-COOH + H_2O$$

or, more generally

$$n\,HO-R'-OH + n\,HOOC-R-COOH \rightarrow$$

$$H + O-R'-OOC-R-CO +_n OH + (2n-1)H_2O$$

(The reader might like to verify for himself that for 5 molecules of diacid and 5 molecules of diol, 5 repeating units of $-O-R'-OOC-R-CO-$ are formed and 9 molecules of water are eliminated).

The early polyesters produced by Carothers were found to have very low softening temperatures. As fibre-forming materials, therefore, they would not withstand ironing and consequently Carothers discarded these polyesters and concentrated his work on polyamides, which led to the development of the nylons. However, in 1942, Whinfield and Dickson produced a polyester which they called polyethylene terephthalate. The introduction of benzene rings into the chain (see opposite) was found to increase chain stiffness and hence softening point, yielding a very useful fibre-forming polyester (see Figure 6, p. 25). In Great Britain, this polyester is known as 'Terylene' and it is the only linear saturated polyester of significant commercial importance.

The starting materials for producing 'Terylene' are ethane-1,2-diol (ethylene glycol) and dimethyl benzene-1,4-dicarboxylate (dimethyl terephthalate); the dimethyl ester is used rather than the free dicarboxylic acid since the former is easier to purify. Ethane-1,2-diol is produced by the catalytic oxidation of ethene followed by hydration:

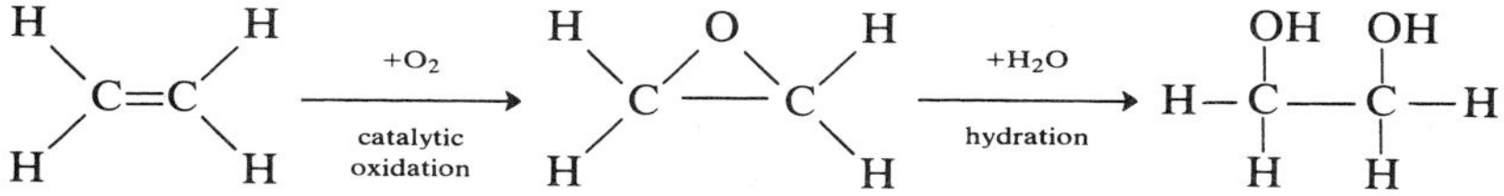

epoxyethane
(ethylene oxide)

The dimethyl benzene-1,4-dicarboxylate is formed by the oxidation of 1,4-dimethylbenzene (*p*-xylene) (using nitric acid, for example) followed by esterification of the carboxylic acid groups:

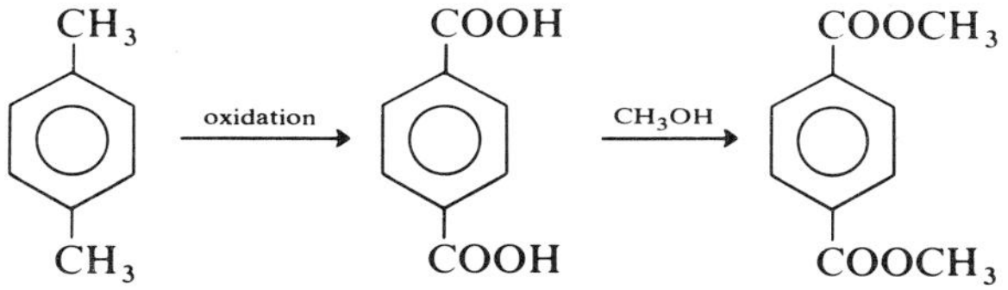

benzene-1,4-
dicarboxylic acid
(terephthalic acid)

The production of the polyester now falls into two further main stages. Firstly, an ester interchange reaction occurs where 1 mol of the dimethyl ester is heated with just over 2 mol of ethane-1,2-diol in the presence of a catalyst; the following reaction occurs:

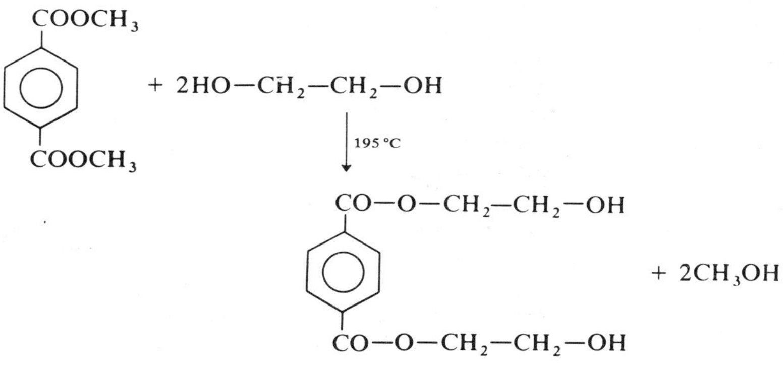

The methanol which is formed is distilled off continuously. In the second stage, which is the polymerisation proper, the product from the first stage is heated to approximately 280 °C at a reduced pressure of about 1 mm Hg. Under these conditions, the ethane-1,2-diol, which is eliminated, is removed continuously from the system, and polymerisation is continued until the desired relative molecular mass is achieved; this stage can be represented as follows:

$$H-O-CH_2-CH_2-O-OC-\left\langle\bigcirc\right\rangle-CO-O-CH_2-CH_2-O-H \; +$$

$$H-O-CH_2-CH_2-O-OC-\left\langle\bigcirc\right\rangle-CO-O-CH_2-CH_2-O-H$$

$$\downarrow$$

$$H\!\vdots\!O-CH_2-CH_2-O-OC-\left\langle\bigcirc\right\rangle-CO\!\vdots\!O-CH_2-CH_2$$

$$H-O-CH_2-CH_2-O\!\vdots\!OC-\left\langle\bigcirc\right\rangle-CO\text{————}O$$

$$+ \; HO-CH_2-CH_2-OH$$

(and so on).

Note that the repeating unit in the 'Terylene' is:

$$\text{—}(O-CH_2-CH_2-O-OC-\left\langle\bigcirc\right\rangle-CO)\text{—}$$

The molten polymer is then extruded from the reaction vessel and on cooling is cut into chips (in extrusion, the hot material is forced through a shaped hole, or die, to form a rod, tube, sheet, ribbon or other continuous length). The polymer is amorphous as it comes from the reaction vessel and will remain essentially amorphous if it is rapidly cooled. It can be crystallised, however, on reheating, and can be converted to filaments by melt spinning (see p. 62). If these filaments are stretched or hot drawn to about four or five times their original length, orientation and crystallinity are induced. These fibres have low water absorption, good wet and dry strengths, and good resistance to degradation by light. They are widely used in fabrics; as any housewife will know, when blended with other fibres, the resulting fabrics are rendered more crease resistant.

Unsaturated polyesters constitute the other group of industrially important polyesters. These materials are frequently used with glass fibre for producing boats, car bodies, etc. These polyesters all have points of unsaturation, or double bonds, along their chain lengths.

Space will not allow a discussion of these materials here, and the reader is referred to the 'Further Reading List' at the end of Chapter 7 if information is desired.

Polyamides

Whilst the amide linkage is to be found in a range of polymers, including naturally occurring proteins (see p. 79), one very important group of polyamides is the *nylons*, which is a general name given to the fibre forming polyamides so well known to us all. Nylons are named according to a numbering system, which refers to the number of carbon atoms in the repeating unit of the polymer. Of the commercially important nylons, we will now look at three—nylon 6,6, nylon 6,10 and nylon 6.

(a) Nylon 6,6. The starting materials for producing nylon 6,6 are hexane-1,6-diamine (hexamethylene diamine) and hexanedioic acid (adipic acid). Hexanedioic acid can be prepared from cyclohexane by the following route:

$$
\begin{array}{ccc}
\underset{\substack{H_2C\text{—}CH_2\\|\qquad|\\H_2C\underset{H_2}{\text{—}C}CH_2}}{\overset{H_2}{C}} & \xrightarrow[\text{air, heat, catalyst}]{O_2} & \underset{\substack{H_2C\text{—}CH\text{—}OH\\|\qquad|\\H_2C\underset{H_2}{\text{—}C}CH_2}}{\overset{H_2}{C}} + \underset{\substack{H_2C\text{—}C{=}O\\|\qquad|\\H_2C\underset{H_2}{\text{—}C}CH_2}}{\overset{H_2}{C}}
\end{array}
$$

cyclohexanol   cyclohexanone

oxidation, using
HNO_3 and catalyst

$$HOOC-(CH_2)_4-COOH$$

Hexane-1,6-diamine can be prepared by the catalytic dehydration of hexanedioic acid in the presence of ammonia to give hexanedinitrile (adiponitrile) followed by hydrogenation:

$$HOOC-(CH_2)_4-COOH \xrightarrow{NH_3}$$

$$NC-(CH_2)_4-CN \xrightarrow[\text{Co or Ni}]{H_2} H_2N-(CH_2)_6-NH_2$$

For fibre production, it is essential that the polymer is of high relative molecular mass, i.e. above 10 000. To achieve this, an equimolar mixture of the diacid and diamine is required (the reader will remember that an excess of one monomer limits the degree of polymerisation—see p. 18). This requirement is helped by the fact that the diacid and diamine form a 1:1 salt (see later), and since this salt has a low solubility in methanol it can be isolated by precipitation. An aqueous salt solution is therefore heated in an autoclave, the temperature eventually reaching about 280 °C—the steam produced displaces air which is present. A small amount of ethanoic acid is also added to limit the relative molecular mass to the desired level (see p. 18). The whole process can be represented by the following equations:

$$n\,H_2N.(CH_2)_6.NH_2 + n\,HOOC.(CH_2)_4.COOH$$

$$\downarrow$$

$$n\,H_3\overset{+}{N}(CH_2)_6\overset{+}{N}H_3.\overset{-}{O}OC(CH_2)_4CO\overset{-}{O}$$
$$\text{nylon salt}$$

$$\downarrow$$

$$H\!+\!NH.(CH_2)_6.NH.CO.(CH_2)_4.CO\!+\!_n OH + (2n-1)H_2O$$

After polymerisation, the molten polymer is extruded as ribbon and is then cut into chips. Nylon 6,6 can be melt spun* and cold drawn to give highly oriented fibres (see Chapter 3) having high strength and good elasticity; they are used for ropes, thread, cord and, of course, for clothing. As a plastic, it has applications in engineering as a substitute for metals in bearings, cams, gears, etc. Nylon 6,6 also has the advantage that it can be injection moulded to close dimensional tolerances.

Note that nylon 6,6 can be prepared in the laboratory using the acid chloride derivative of hexanedioic acid, $ClOC.(CH_2)_4.COCl$, instead of the free acid—the only difference is that HCl is now the small molecule eliminated, instead of water. The details are given in experiment **4**, Chapter 7, and the reader may note that nylon 6,10 (see below) can also be prepared using a similar method.

(*b*) *Nylon 6,10.* This nylon is produced by the polymerisation of hexane-1,6-diamine with decanedioic acid (sebacic acid) (see Figure 24(a)). The monomers, when mixed, form a nylon 6,10 salt, and the polymerisation is similar to that described above for the salt of nylon 6,6. The repeating unit for the nylon 6,10 polymer is shown in Figure 24(b).

$$HOOC-(CH_2)_8-COOH \qquad +NH.(CH_2)_6.NH.CO.(CH_2)_8.CO+$$
$$(a) \qquad\qquad\qquad\qquad (b)$$

Figure 24 Structures of (a) decanedioic acid and (b) the repeating unit for nylon 6,10.

---

* Melt spinning involves pumping molten polymer at a constant rate under high pressure through a plate, called a spinneret, containing a number of holes. The liquid polymer 'jets' emerge downward from the face of the spinneret, usually into the air. As they solidify, they are brought together to form a thread and are then wound up on bobbins. The fibres can now be oriented by drawing.

Nylon 6,10 has not been used as a textile fibre, but is used for monofilaments for brushes and bristles. The additional $-CH_2$ groups in the chains result in a lower melting point and water absorption than for nylon 6,6, and it retains its stiffness and mechanical properties better when wet than does nylon 6,6.

(c) *Nylon 6*. This polymer is prepared from cyclohexanol via the following route:

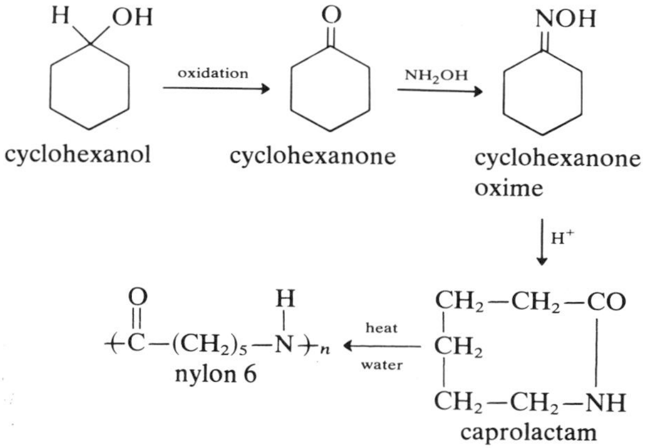

cyclohexanol   cyclohexanone   cyclohexanone oxime

nylon 6   caprolactam

Polymerisation is carried out by the addition of water to open the ring, followed by the removal of water at high temperature. A linear polymer is produced which is in equilibrium with about 10 per cent of monomer—the latter is removed by washing with water. The polymer can be melt spun and cold drawn to give fibres of high strength and good elastic properties.

Polyurethanes

The isocyanate group, $-NCO$, is a very reactive group and it will form urethanes with alcohols:

$$R.NCO + R'OH \rightarrow R.NH.COO.R'$$

It is clear that if a di- or polyisocyanate is reacted with a di- or polyol (polyhydric compound) then a polyurethane will be formed:

$$OCN-R-NCO + HO-R'-OH \rightarrow OCN-R-NH-CO-O-R'-OH$$

further reaction with monomers

$$(CO-NH-R-NH-CO-O-R'-O)_n$$

Like polyamides, polyurethanes can take part in hydrogen bonding.

The first commercial attempt to produce polyurethanes was by Bayer in Germany who produced a polymer from hexane-1,6-diisocyanate (hexamethylene diisocyanate) and butane-1,4-diol (1,4-butanediol); the repeating unit has the structure:

$$
+\!\overset{\displaystyle O}{\overset{\displaystyle \|}{C}}-\overset{\displaystyle H}{\overset{\displaystyle |}{N}}+\!CH_2)_6-\overset{\displaystyle H}{\overset{\displaystyle |}{N}}-\overset{\displaystyle O}{\overset{\displaystyle \|}{C}}-O+\!CH_2)_4-O+\!
$$

The polymer was found to have very similar properties to nylon, but has achieved little commercial importance because of its lower melting point and problems with dyeing. Since then, however, major advances have been made in polyurethane chemistry resulting in polyurethane foams, elastomers, surface coatings, fibres and adhesives.

Polyurethane foams can be formed by simultaneously producing a polyurethane polymer and a gas. If these processes are correctly balanced, bubbles of the gas are trapped in the polymer matrix as it is formed, hence producing the foam. Both flexible and rigid foams can be produced—slightly cross-linked foams are flexible whilst highly cross-linked products are rigid. In forming a flexible foam, two main reactions are proceeding simultaneously:

diisocyanate + polyol → polyurethane

diisocyanate + water → carbon dioxide

The second reaction, therefore, produces the blowing agent, which is carbon dioxide. Flexible foams are based either on polyesters or polyethers (see later). In other words, the polyols are in fact low relative molecular mass polyesters or polyethers bearing hydroxyl groups at their ends. Polyether foams now account for the majority of flexible foams produced.

For rigid foams, the same essential principles are used as for flexible foams. The main difference between the two products is in the degree of cross-linking. Rigid foams are highly cross-linked and this is achieved by using relatively low relative molecular mass polyols, which are again mainly polyethers rather than polyesters. Also, the blowing agents can differ—carbon dioxide (produced by the diisocyanate–water reaction) can also be used to make rigid foams, but usually low-boiling, inert, halogenated alkanes, such as CCl_3F, are used. These liquids take no part in the chemical reaction, but are simply volatilised by the heat of polymerisation, hence expanding the foam. Flexible foams are used in upholstery, and this is now a major industry. Rigid foams are widely used as heat insulation materials.

Polyurethanes, as mentioned above, are also used in the production of elastomers. They possess good mechanical properties, having high tear and abrasion resistance. They are unfortunately rather highly priced, however, and this has restricted their use somewhat. In the field of surface coatings, many successful polyurethane paint and varnish preparations have now deservedly established themselves on the market because of their good weather and abrasion resistance generally.

Polyethers

An important group of polyethers is the epoxy resins or epoxides. The most common of these resins are produced by the reaction of epoxy compounds with dihydric phenols, followed by cross-linking with di- or polyfunctional amines, acids or anhydrides. A typical example is the polymer made from 1-chloro-2,3-epoxypropane (epichlorhydrin) and a dihydric phenol known by the trivial name of diphenylol propane:

$$n\,H_2C\overset{O}{\overset{/\backslash}{-}}CH-CH_2Cl \;+\; n\,HO-\langle\bigcirc\rangle-\underset{CH_3}{\overset{CH_3}{\underset{|}{\overset{|}{C}}}}-\langle\bigcirc\rangle-OH$$

$$\text{NaOH}\;\Big\downarrow\;-\text{HCl}$$

$$+CH_2-\underset{}{\overset{OH}{\overset{|}{CH}}}-CH_2-O-\langle\bigcirc\rangle-\underset{CH_3}{\overset{CH_3}{\underset{|}{\overset{|}{C}}}}-\langle\bigcirc\rangle-O+_n$$

The epoxy resins have wide use as surface coatings, giving films which are highly resistant to chemical attack. They are used in a whole range of cold-setting and thermosetting adhesives which can be used to join a wide range of different materials. In the retail packs of epoxy adhesives there are two tubes provided, one containing the resin and the other containing the hardener (which consists of one of the cross-linking agents mentioned earlier). When mixed, the low relative molecular mass resin molecules are linked together to form a tough, cross-linked material with a high relative molecular mass. The high strengths of epoxy adhesives, such as 'Araldite', have led to their being used for bonding metals instead of soldering joints.

Before leaving this section, it should be mentioned that polyethers can also be formed by addition polymerisation. For example, poly(epoxyethane) (polyethylene oxide or 'Polyox'), having the repeating unit $+CH_2-CH_2-O+$, can be produced by an anionic mechanism, and polymers having relative molecular masses of up to 5 million can be formed. In recent years, one area of interesting research involving 'Polyox' has been in the field of drag reduction.

A successful application of this phenomenon has been in fire fighting in the USA where drag reduction in water has increased 'jet throw', hence enabling fires in tall buildings to be more efficiently combated.

Phenoplasts

Resins made from phenols and aldehydes constitute the group known as *phenolics* or *phenoplasts*. Phenols will react with aldehydes to give condensation products provided that the 2- (*ortho*) and 4- (*para*) positions relative to the phenolic hydroxyl group are available for reaction—the reader will remember that the 2- and 4- positions in phenol are activated because the following electron delocalisation increases the electron density at these positions:

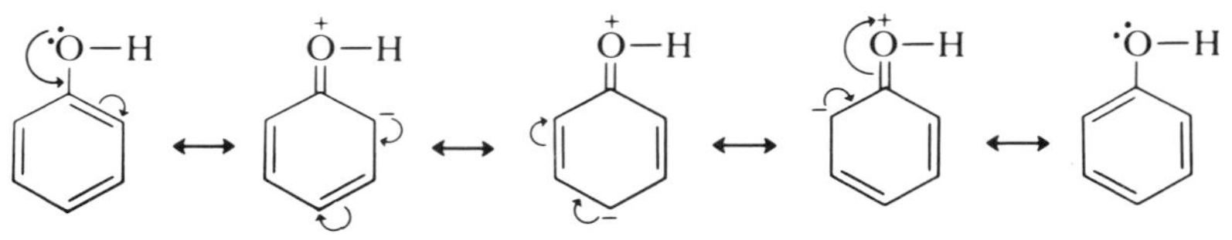

The aldehyde used almost exclusively in the industrial production of phenoplasts is methanal (formaldehyde). The reaction is catalysed by acids or bases, and the product formed depends on both the type of catalyst used and the phenol–methanal molar ratio (known as the P/F ratio, from the old name phenol–formaldehyde). The reaction between the phenol and the methanal results in the introduction of $-CH_2OH$ groups into the 2- and 4- positions of the phenol. Obviously, various phenol–alcohols can be formed, depending on the P/F ratio used (Figure 25).

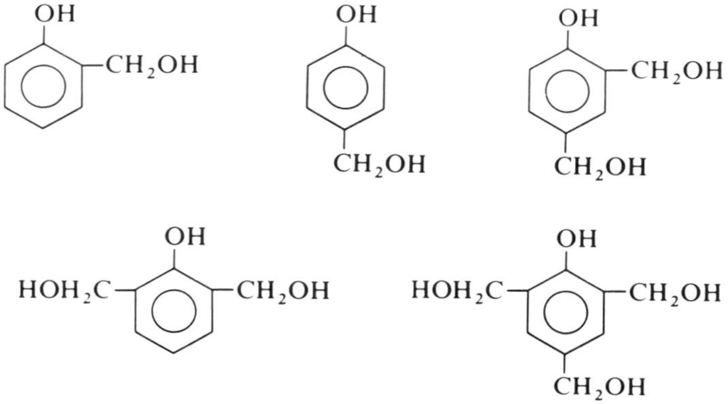

Figure 25 The various phenol–alcohols which can be formed using different P/F ratios.

When an alkaline catalyst is used with a P/F ratio of less than unity (i.e. an excess of methanal), phenol–alcohols can condense with each other to form short chain compounds called *resols*—these are soluble in the reaction mixture. An example of resol formation is shown in Figure 26.

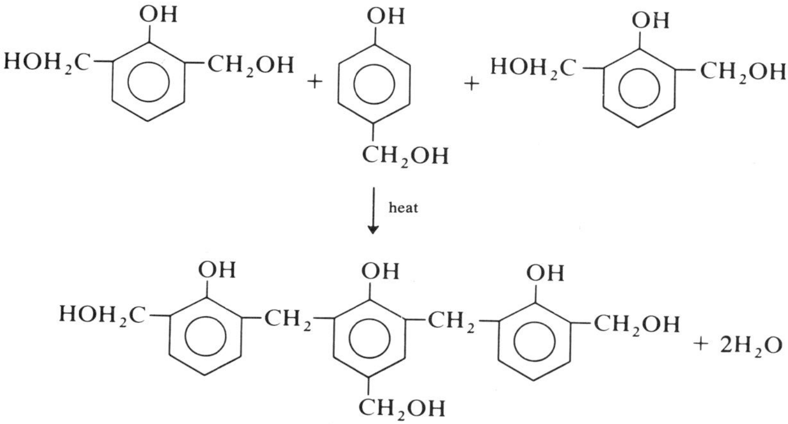

Figure 26 Scheme depicting resol formation.

Under these conditions, the reaction between phenol and methanal can be separated into three stages (as originally distinguished by Baekeland).

Stage A. The formation of resols, as above, which are soluble in the reaction mixture.

Stage B. Growth to form larger chains known as *resistols*—these are still soluble in various solvents so are not highly cross-linked.

Stage C. The development of extensive cross-linking to form a hard, infusible and completely insoluble resin called a *resit*.

Studies have shown that of the various cross-linking reactions which can occur, the most important ones are those involving the formation of $-CH_2-$ bridges (known as 'methylene' bridges—Figure 27) and $-CH_2-O-CH_2-$ bridges (known as ether bridges—Figure 28).

An entirely different product is obtained if an acid catalyst is used with a P/F ratio greater than unity (i.e. an excess of phenol). The products are called *novolaks*, and are fusible and soluble in solvents for later use. They are short chain linear polymers, as shown in Figure 29, having a random arrangement of 2- and 4- (*ortho* and *para*) linkages.

The difference between novolaks and resols is that with the former polymers, the smaller amount of methanal means that there are no

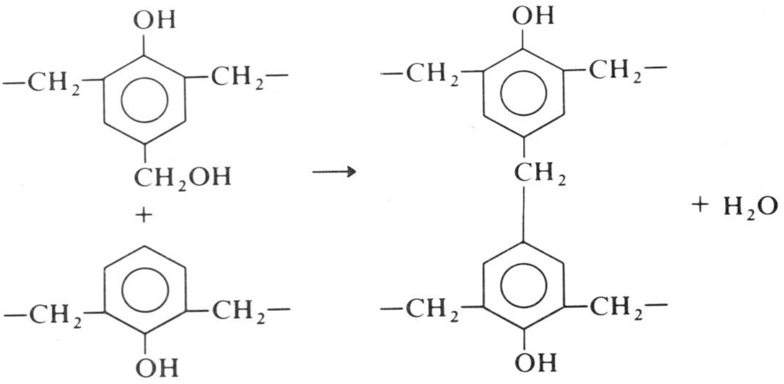

Figure 27 Formation of 'methylene' bridges in phenolic resins.

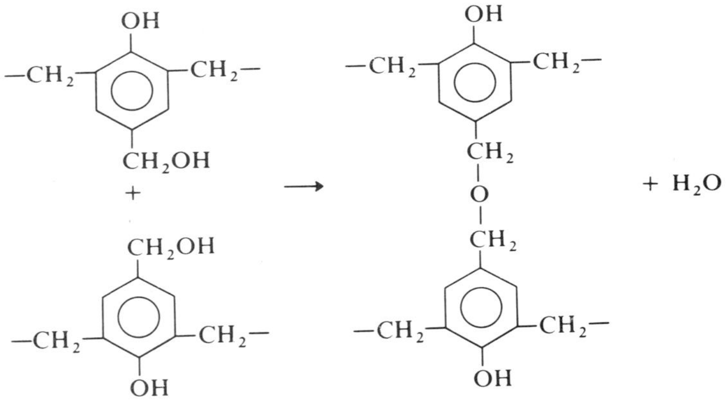

Figure 28 Formation of ether bridges in phenolic resins.

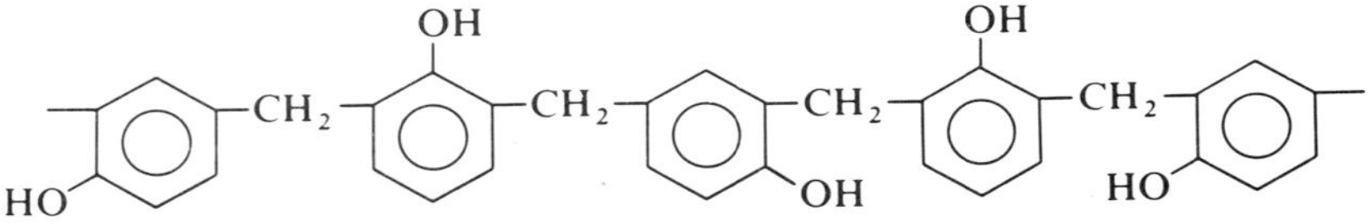

Figure 29 Section of a novolak polymer.

free $-CH_2OH$ groups present (as there are in resols); hence, cross-linking cannot occur. Cross-linking of novolaks can be carried out if $-CH_2OH$ groups are introduced, and this can be done by heating the novolak with an alkaline solution of methanal.

It was in the early 1900s that Baekeland first developed a thermo-setting phenol–methanal resin, which was termed 'Bakelite', and since then this area of polymer chemistry has become a major industry.

Phenolic resins when fully cured are, generally speaking, resistant to high temperature, solvents and chemicals, and are good electrical insulators. Plastics made from them are hard, rigid and generally dark in colour. They find uses in a wide range of goods, including electrical switches and fittings, ashtrays, saucepan and door handles, car fittings, etc. In addition to moulding and casting, phenol–methanal resins are used as adhesives (for bonding plywood, for example) and for producing laminates.

A simple preparation of a phenol–methanal resin is given in experiment **6**, Chapter 7.

Aminoplasts

Aminoplasts are produced by the condensation of methanal with organic amino compounds such as urea (or carbamide) and melamine. Urea, which is prepared by the reaction of carbon dioxide with ammonia, reacts with methanal under slightly alkaline or neutral conditions to give compounds known by the trivial names of monomethylolurea and dimethylolurea; products obviously depend on the ratio of the reactants:

Under acid conditions, 'bisamides' are formed by the linking of pairs of urea molecules with methanal:

Methylol derivatives are then formed by addition of methanal to the 'bisamide':

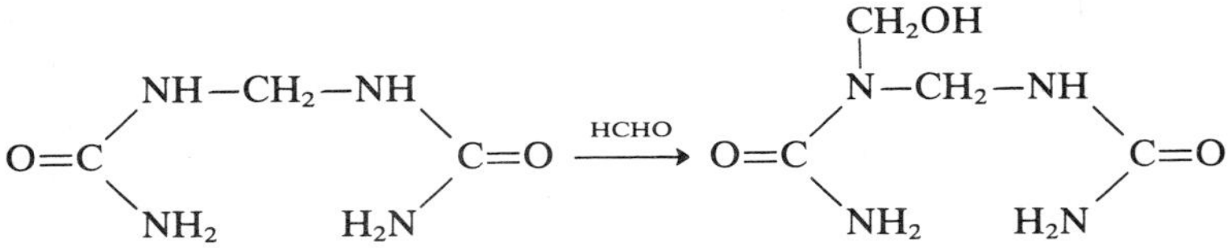

$$O=C\underset{NH_2}{\overset{NH-CH_2-NH}{<}}C=O \quad \xrightarrow{\text{HCHO}} \quad O=C\underset{NH_2}{\overset{N-CH_2-NH}{<}}C=O$$

with the upper left group bearing CH_2OH on the nitrogen.

(The methylol groups begin to cross-link rapidly under these conditions.)

The mechanism of the formation of linear polymers, and subsequent cross-linking, is not fully understood. Monomethylolurea, when heated in alkaline conditions, gives water-soluble condensation products—on acidifying and heating, the following reaction may occur:

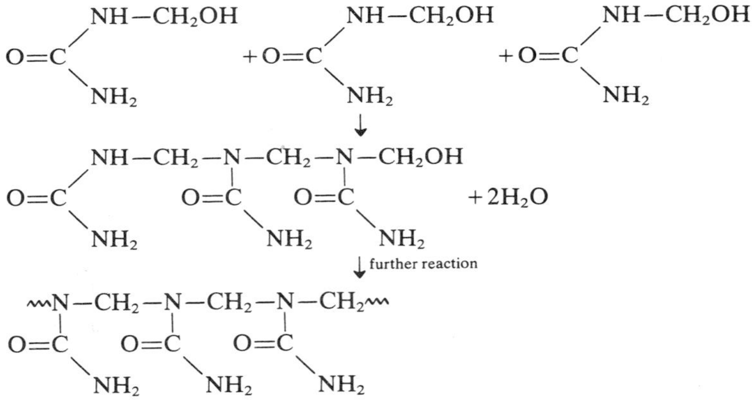

$$O=C\underset{NH_2}{\overset{NH-CH_2OH}{<}} + O=C\underset{NH_2}{\overset{NH-CH_2OH}{<}} + O=C\underset{NH_2}{\overset{NH-CH_2OH}{<}}$$

$$\downarrow$$

$$O=C\underset{NH_2}{\overset{NH-CH_2-N-CH_2-N-CH_2OH}{<}} \quad + 2H_2O$$

with $O=C$ and NH_2 on the bridging nitrogens.

$$\downarrow \text{ further reaction}$$

$$\sim\sim N-CH_2-N-CH_2-N-CH_2\sim\sim$$

with $O=C$ and NH_2 groups on each nitrogen.

It is clear that with dimethylolurea, cross-linking can occur and, in fact, insoluble infusible products are obtained. Urea–methanal (urea-formaldehyde or U/F) resins are light in colour and, although they are not as resistant to hot water as phenolics, they have good resistance to the usual organic solvents, oils and greases. Familiar applications include light switches and fittings, bathroom fixtures, tumblers, dishes and similar moulded articles. They also find considerable use as adhesives, being employed in the production of plywood, and in laminating and veneering.

Melamine has the structure:

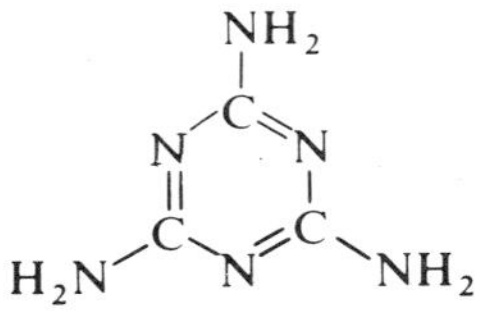

$$
\begin{array}{c}
NH_2 \\
| \\
C \\
N \diagdown \diagup N \\
\| \quad \| \\
C \quad C \\
H_2N \diagup N \diagdown NH_2
\end{array}
$$

Like urea, it reacts with methanal under neutral or alkaline conditions to form methylol melamines. On acidifying or further heating, cross-linked resins (known as M/F resins) are formed by a rather complex process. Although M/F resins are more expensive than U/F resins, they have some advantages over the latter including better heat and chemical resistance, hardness and low water absorption. They are used for making tableware and decorative laminates for table-tops.

Silicones

Although the chemistry of carbon is dominated by its propensity to form long stable chains and rings, silicon, whilst being in the same group in the Periodic Table, does not form stable $-Si-Si-$ chains of any great length. However, chains of the type $-Si-O-Si-O-Si-O-$ are very stable, and form the backbone or skeleton of a group of polymers known as the *silicones*.

The starting point for the manufacture of silicone polymers is the production of organo-silicon chlorides. For example, in what is known as the *direct process*, chloromethane (methyl chloride) is passed through a finely divided mixture of silicon and copper (catalyst) at about 280 °C. The composition of the product cannot be easily regulated by manipulating the reaction conditions; separation of the mixture of products obtained, therefore, is accomplished by very careful fractional distillation. The main reaction which occurs can be represented as follows:

$$2CH_3Cl + Si \xrightarrow[\text{Cu}]{280\,°C} (CH_3)_2SiCl_2$$

Small amounts of methyltrichlorosilane, CH_3SiCl_3, trimethyl-chlorosilane, $(CH_3)_3SiCl$, and other products are also obtained.

These chloro compounds are readily hydrolysed by water:

$$\underset{\underset{CH_3}{|}}{\overset{\overset{CH_3}{|}}{Cl-Si-Cl}} + 2H_2O \rightarrow \underset{\underset{CH_3}{|}}{\overset{\overset{CH_3}{|}}{HO-Si-OH}} + 2HCl$$

The hydroxyl groups of these hydrolysis products are unstable, however, and react with each other, eliminating water, to form $-Si-O-Si-$ linkages. In the case of dimethyldichlorosilane, $(CH_3)_2SiCl_2$, hydrolysis then leads to the formation of a linear polymer:

$$HO-\underset{\underset{CH_3}{|}}{\overset{\overset{CH_3}{|}}{Si}}-OH + HO-\underset{\underset{CH_3}{|}}{\overset{\overset{CH_3}{|}}{Si}}-OH + HO-\underset{\underset{CH_3}{|}}{\overset{\overset{CH_3}{|}}{Si}}-OH$$

$$\downarrow$$

$$HO-\underset{\underset{CH_3}{|}}{\overset{\overset{CH_3}{|}}{Si}}-O-\underset{\underset{CH_3}{|}}{\overset{\overset{CH_3}{|}}{Si}}-O-\underset{\underset{CH_3}{|}}{\overset{\overset{CH_3}{|}}{Si}}-OH + 2H_2O$$

$$\downarrow \text{further reaction}$$

$$HO\left(\underset{\underset{CH_3}{|}}{\overset{\overset{CH_3}{|}}{Si}}-O\right)_nH + (n-1)H_2O$$

whereas in the case of trimethylchlorosilane, hydrolysis leads only to the production of a dimer since this compound is monofunctional. Compounds of the type R_3SiCl, where R is an alkyl or aryl group, are therefore used as 'end-stoppers' to regulate the chain lengths of silicone polymers produced from the dichloro- derivative; obviously, the more R_3SiCl which is used, the shorter will be the average chain length. It is now easy to see that if methyltrichlorosilane is hydrolysed, subsequent condensation between hydroxyl groups will lead to the production of three-dimensional cross-linked structures:

$$CH_3-\underset{\underset{Cl}{|}}{\overset{\overset{Cl}{|}}{Si}}-Cl + 3H_2O \rightarrow CH_3-\underset{\underset{OH}{|}}{\overset{\overset{OH}{|}}{Si}}-OH + 3HCl$$

$$-H_2O \downarrow \text{condensation}$$

$$CH_3-\underset{\underset{O}{|}}{\overset{\overset{O}{|}}{Si}}-O-\underset{\underset{O}{|}}{\overset{\overset{CH_3}{|}}{Si}}-O-\underset{\underset{O}{|}}{\overset{\overset{O}{|}}{Si}}-CH_3$$

$$CH_3-\underset{\underset{O}{|}}{\overset{\overset{O}{|}}{Si}}-O-\underset{\underset{CH_3}{|}}{\overset{\overset{O}{|}}{Si}}-O-\underset{\underset{O}{|}}{\overset{\overset{O}{|}}{Si}}-CH_3$$

Silicone polymers fall into three groups—the fluids, the elastomers and the resins. Silicone fluids are colourless liquids which can be prepared to give a range of viscosities. They are essentially linear polymers of rather low relative molecular mass, and the most common are dimethyl silicone fluids prepared from the hydrolysis of dimethyl-dichlorosilane in acid solution. Their low volatility, high flash point and inertness make them excellent lubricants under extreme conditions of heat and low pressure. With suitable fillers, fluids can be converted to lubricating greases (the properties of low flammability and volatility also make silicone fluids very useful for heating baths in the laboratory). The most important application of these fluids, however, is in the development of products which can be applied to fabrics, leather and paper to render them water repellent.

The silicone elastomers are based on linear polymers, analogous to the fluids, but of very high relative molecular mass. After mixing with various fillers, curing is carried out by heating the silicone with an organic peroxide to yield an elastic product. Curing is thought to occur via cross-linking between the methyl groups of adjacent polymer chains, forming $-Si-CH_2-CH_2-Si-$ linkages. A possible scheme could be as follows:

$$
\begin{array}{ccc}
\begin{array}{c}
CH_3 \\
| \\
\sim\!Si-O\!\sim \\
| \\
CH_3 \\
\\
CH_3 \\
| \\
\sim\!Si-O\!\sim \\
| \\
CH_3
\end{array}
&
\xrightarrow[]{2RO\cdot}
\begin{array}{c}
CH_3 \\
| \\
\sim\!Si-O\!\sim \\
| \\
\cdot CH_2 \\
\\
\cdot CH_2 \\
| \\
\sim\!Si-O\!\sim \\
| \\
CH_3
\end{array}
\;+2ROH \rightarrow
&
\begin{array}{c}
CH_3 \\
| \\
\sim\!Si-O\!\sim \\
| \\
CH_2 \\
| \\
CH_2 \\
| \\
\sim\!Si-O\!\sim \\
| \\
CH_3
\end{array}
\end{array}
$$

Silicone elastomers show remarkable stability at high temperatures whilst retaining their elastic properties at low temperatures (the range can be as large as $-90\,°C$ to $250\,°C$). They are also resistant to many chemicals, oils and weathering, but have generally poor resistance to hydrocarbon solvents. Their heat resistance, however, makes them useful as gaskets and seals in ovens and engines, and for tubing which needs to be repeatedly sterilised by autoclaving (hence giving them a variety of surgical applications).

Finally, silicone resins have a three-dimensional branched-chain structure. They are usually prepared by the hydrolysis of mixtures of the di- and trichlorosilanes, and obviously the degree of cross-linking will be governed by the molar ratio of these two compounds (the greater the amount of trichloro- derivative, the greater the amount of cross-linking). Resins can vary, therefore, from very flexible materials to hard, brittle glassy materials. As with the fluids, resins are highly water-repellent and resistant to most aqueous chemical agencies; their resistance to many organic solvents is poor, however. Resins are used in water-repellent treatments for brickwork, in varnishes, paints, 'non-stick' surfaces, and for impregnating glass fibre to make laminates.

Polysulphides

1,2-Dichloroethane and sodium polysulphide, when heated together, form a polysulphide polymer. This can be represented by the following general equation:

$$n\,CH_2Cl.CH_2Cl + n\,Na_2S_x \rightarrow \left(CH_2.CH_2.S_x\right)_n + 2n\,NaCl$$

Sodium polysulphide can be produced by heating sulphur in an aqueous solution of sodium hydroxide; this process can be written as follows:

$$6NaOH + (2x+1)S \rightarrow 2Na_2S_x + 3H_2O + Na_2SO_3$$

These polymers, called 'thiokols', have now been prepared using a variety of dichlorides. Vulcanisation can be carried out by heating the products with metallic oxides. The resultant elastomers, whilst mechanically inferior to rubber, have very high resistance to hydrocarbon oils and solvents; this makes them very useful for the manufacture of gaskets, petrol hoses, etc. A simple preparation of a thiokol is given in experiment **8**, Chapter 7.

Naturally Occurring High Polymers $\quad$ **6**

Living matter, both animal and vegetable, consists largely of polymeric material. In addition, many minerals such as the silicates can also be classified as polymeric. When we realise that proteins, for example, are polycondensates of amino acids, it is evident that the very basis of life is dominated by polymers. Throughout this chapter, therefore, we shall look very briefly at selected examples of naturally occurring polymers and their importance to us.

Polysaccharides

Carbohydrates are compounds consisting of only carbon, hydrogen and oxygen; the majority can be expressed by the general formula $C_x(H_2O)_y$. We can classify carbohydrates into two principal groups known as the *sugars* and the *polysaccharides*. The sugars are usually crystalline, water-soluble materials having exact relative molecular masses, and can be divided into mono-, di-, trisaccharides, etc. The two best known monosaccharides are glucose and fructose, both of molecular formula $C_6H_{12}O_6$, and both referred to as *hexoses* because they contain six carbon atoms.

Polysaccharides are macromolecular structures built up from many monosaccharide units joined together by the loss of water. Hence, polysaccharides can be broken down on hydrolysis to form monosaccharides. Cellulose, starch and glycogen are amongst the better known polysaccharides.

(a) Cellulose

The fibrous tissue in the cell walls of plants contains the polysaccharide cellulose. It is, therefore, a very abundant and widely distributed natural polymer, and millions of tonnes of cellulose are used every year in its three main outlets, i.e. lumber, textiles and paper. Its principal source is wood. Most woods contain up to about 50 per cent of cellulose, together with other constituents such as lignin. Separation of the cellulose from the wood involves digesting with an aqueous solution of sulphur dioxide and a hydrogensulphite (bisulphite) in the sulphite process, or with an aqueous solution of sodium hydroxide and sodium sulphide in the sulphate (or Kraft) process (the latter process is called the sulphate process because sodium sulphate is used as the source of sodium sulphide). In both processes, the lignin is taken into solution, enabling the cellulose to be recovered. The other important source of cellulose is cotton, which is almost pure cellulose. Extraction is carried out by treating with aqueous sodium hydroxide under pressure followed by bleaching with gaseous chlorine or calcium chlorate(I) (hypochlorite).

The molecular formula of cellulose is $(C_6H_{10}O_5)_n$ where n can be several thousand. It is difficult to measure the relative molecular mass of cellulose because (i) there are very few solvents for the polymer (see p. 77), (ii) cellulose is very prone to degradation during the course of any work or processing, and (iii) there is the complication of using cellulose from different sources. It is often found preferable to fully nitrate the cellulose by a non-degradative method and hence obtain the relative molecular mass of the cellulose from that of the nitrate. In this way, values of about one million have been determined for cellulose from cotton.

Cellulose is built up of chains of β-1,4 linked glucose units, and to understand this terminology we first need to look at the structure of glucose itself. Glucose has the molecular formula $C_6H_{12}O_6$ and its structure is shown in Figure 30. In other words, we have represented the structure of glucose both as a straight chain and as a ring structure. The ring structure can be formed as a result of internal hemiacetal formation (Figure 31). However, a closer study of this mechanism shows that two possible configurations for the resulting ring are possible, depending on which way the —OH group at carbon atom 1 (C1) is pointing.

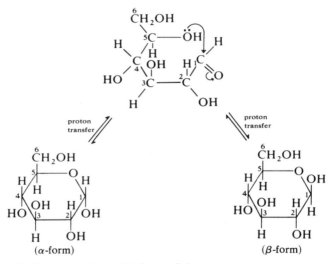

$$
\begin{array}{cc}
\text{CHO} & \text{CHOH} \\
| & | \\
\text{CHOH} & \text{CHOH} \\
| & | \\
\text{CHOH} & \text{CHOH} \\
| & | \\
\text{CHOH} & \text{CHOH} \\
| & | \\
\text{CHOH} & \text{CH} \\
| & | \\
\text{CH}_2\text{OH} & \text{CH}_2\text{OH} \\
(a) & (b)
\end{array}
$$

Figure 30 Diagram showing the (a) open chain and (b) ring structures of glucose.

Figure 31 Formation of α- and β-forms of glucose.

When the —OH group at C1 is pointing down, the glucose takes the α-form (and hence the β-form if the —OH at C1 is pointing upwards). In solution, the two forms are in equilibrium with each other; in fact, because glucose also shows the reducing properties typical of aldehydes (e.g. reactions with Tollens's reagent and Fehling's solution), this is evidence for the presence of small amounts of the open or straight chain structure as well.

We are now in a position to understand what is meant by a β-1,4 linkage. As stated earlier in the chapter, polysaccharides are built up from many monosaccharide units joined together by the loss of water, and this results in a *glycosidic* linkage (which is an oxygen bridge). The glycosidic linkage in cellulose between the C1 of one unit and the C4 of the next unit is shown in Figure 32, and it is a β-glycosidic linkage.

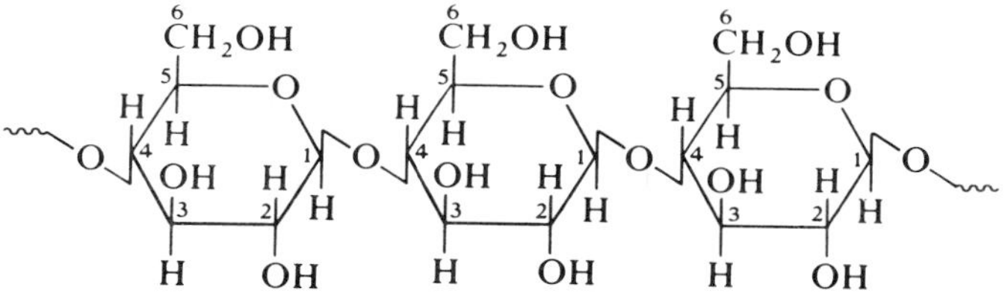

Figure 32 The β-glycosidic linkage in cellulose.

It is also now evident why cellulose is classed as non-reducing, because the point of linkage is at the reducing C1 carbon atom (see also Figure 30(a)).

On inspecting the structure of cellulose, one could be forgiven for thinking that it might have a high water solubility since its high hydroxyl content would naturally suggest hydrogen bonding with water (i.e. high solute–solvent interactions). However, this is not the case, and cellulose is not only insoluble in water but in most other reagents as well. This is because of the stiffness of the chains and the high interchain forces due to hydrogen bonding between hydroxyl groups on adjacent chains. These factors are also thought to be responsible for the often highly crystalline nature of cellulose fibres. If hydrogen bonding (and hence interchain forces) is to be reduced or eliminated, then the hydroxyl groups of the cellulose must be replaced partially or wholly by, for example, esterification. This can be done and the resulting esters are, in fact, soluble in a number of solvents. Cellulose has also been found to be soluble in ammoniacal solutions of copper(II)hydroxide; complex formation involving the hydroxyl groups of the cellulose, the Cu^{2+} ions and the ammonia has been suggested as one explanation for this phenomenon.

The amount of cellulose which occurs in a physical form which is directly usable for fibres, etc., is somewhat limited. It is common practice, therefore, to process a solution of a cellulose derivative and then having manipulated the polymer into the required shape (e.g. fibre or film) to remove the modifying groups to regenerate unmodified cellulose. Such a material is therefore known as *regen-erated cellulose*. Fibres manufactured from cellulose are called *rayon*, and their production can make use of the above technique. For example, an early process for producing regenerated cellulose fibre involved treating cellulose with the solution of ammoniacal copper(II)hydroxide mentioned above; the resulting solution was then forced through spinnerets into an acid bath to regenerate the cellulose as long filaments. An alternative method of regeneration is to dissolve the cellulose in a solution of sodium hydroxide and carbon disulphide (Figure 33). The resulting solution, called *viscose*, is then forced through a spinneret into an acid solution, when the cellulose is regenerated as a fibre which can be further processed. The product is known as *viscose rayon* and is now a major commercial textile fibre. Alternatively, if the viscose solution is forced through a thin slit into the acid solution instead of through the holes of the spinneret, the cellulose is regenerated as a thin sheet, and when further processed this is sold as *cellophane*. This latter reaction can be represented by the scheme in Figure 33.

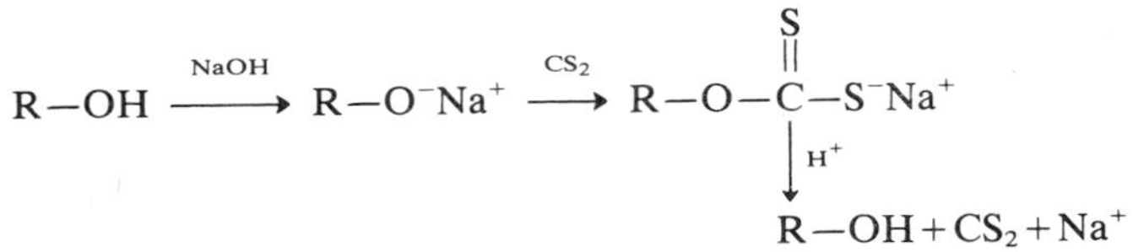

$$R-OH \xrightarrow{\text{NaOH}} R-O^-Na^+ \xrightarrow{\text{CS}_2} R-O-\overset{\displaystyle S}{\underset{\displaystyle H^+}{\overset{\|}{C}}}-S^-Na^+$$

$$\downarrow H^+$$

$$R-OH+CS_2+Na^+$$

Figure 33 Regeneration of cellulose using the sodium hydroxide and carbon disulphide system, where R represents the cellulose chain.

Cellulose ethanoate (cellulose acetate) can be prepared by heating cellulose with ethanoic anhydride (acetic anhydride) and ethanoic acid (acetic acid) in the presence of sulphuric acid. It is used for making fibres, films and lacquers.

(*b*) *Starch and glycogen*

Starch is again a naturally occurring, widely distributed polymer of molecular formula $(C_6H_{10}O_5)_n$. It is present in wheat, barley, rice, potatoes and all green plants. Its main sources are grains, such as corn. In the extraction of starch, the plant material is ground with water, and the resulting slurry is then filtered to remove coarse tissue fragments, hence leaving a suspension of starch granules. The granules are then collected by centrifuging this filtrate.

Starch contains two types of polymers which differ in structure and relative molecular mass, and are known as *amylose* and *amylopectin*. Amylose, which accounts for about 20–25 per cent of natural starch, is made up of glucose units joined by α-1,4 linkages (Figure 34); its relative molecular mass varies widely depending on the source (as does the relative molecular mass of amylopectin).

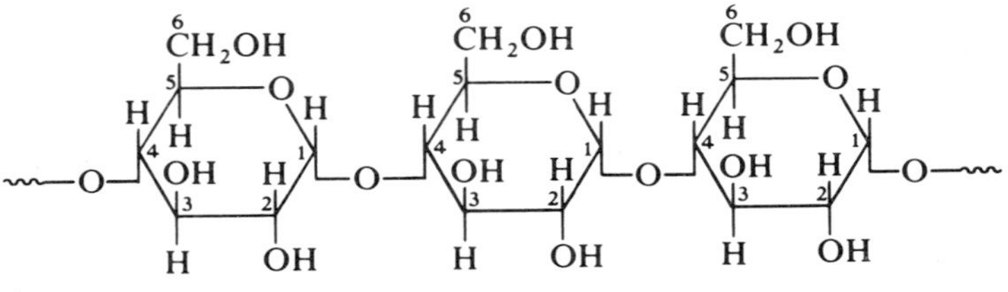

Figure 34 Structure of amylose.

The other component of starch, amylopectin, is a branched-chain polymer, having α-1,6 in addition to α-1,4 glycosidic linkages (Figure 35).

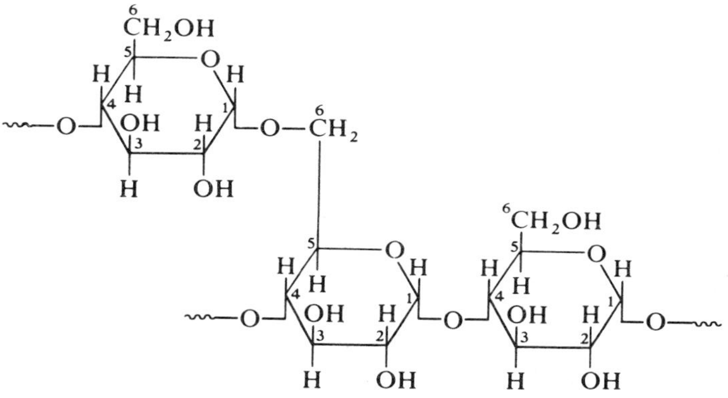

Figure 35 Structure of amylopectin.

Again (as for cellulose), it is obvious from these structures that starch will be non-reducing. It does, however, form a characteristic blue-black colour in the presence of iodine, hence making starch solution a useful indicator in volumetric analysis.

Starch is used extensively in the paper industry and for paper adhesives. A considerable amount is used by the food industry in foodstuffs, and much starch is hydrolysed to give glucose. Starch is also used for producing sizes for paper and textiles, and for fermentation to alcohol.

Glycogen is an important energy store in the body and is found mainly in muscle and liver. It is a highly branched polymer having a structure similar to that of amylopectin.

Proteins

We have just seen that cellulose is the main constituent of the cell walls of plants. Proteins, on the other hand, are the main constituents of all living cells in animals and plants. Unlike carbohydrates, proteins contain nitrogen as well as carbon, hydrogen and oxygen; sulphur is often present as well. Proteins are derived from α-amino acids (an α-amino acid is one which contains the —COOH and the —NH$_2$ groups attached to the same carbon atom) which are joined together via the amino group of one acid molecule and the carboxyl group of the next, accompanied by the elimination of water. For example, consider the two α-amino acids glycine and alanine (see Table 3) joined together in the above manner:

$$H_2N-CH_2-CO-NH-\overset{\overset{\displaystyle CH_3}{|}}{CH}-COOH$$

This resulting molecule is called a *dipeptide*, and the —CO.NH— link is known as the *peptide* link. Three α-amino acids linked together give rise to a *tripeptide*, and so on to finally give a *polypeptide*. Proteins are high polymeric polypeptides; in other words, if only a relatively small number of amino acid units are condensed together, we refer to the chain as a polypeptide, whereas a very high relative molecular mass polymer is known as a protein. The division between the two is sometimes put at relative molecular mass 10 000, but the distinction is an imprecise, rather arbitrary one. The general structure of a polypeptide or protein, therefore, can be represented as in Figure 36, and the nature of some of the R groups can be seen in Table 3.

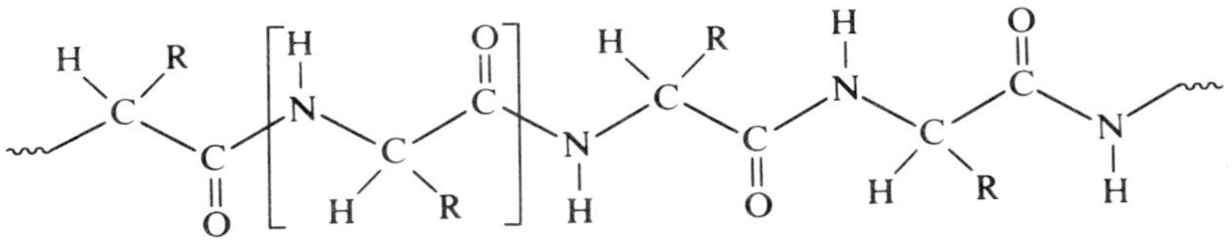

Figure 36 A polypeptide or protein chain (for nature of R groups, see Table 3).

Reference to Figure 36 shows that the addition of water to the repeating unit shown in brackets gives rise to an α-amino acid. It is not surprising, therefore, that proteins can be hydrolysed back to α-amino acids, and acid hydrolysis of proteins is a technique used in the laboratory for determining protein structure (see later).

Table 3 Some α-amino acids obtained from proteins (names are given using old nomenclature).

H_2N-CH_2-COOH glycine	$H_2N-\overset{\overset{\displaystyle CH_3}{|}}{^*CH}-COOH$ alanine	$H_2N-\overset{\overset{\displaystyle SH}{\overset{\displaystyle |}{\overset{\displaystyle CH_2}{|}}}}{^*CH}-COOH$ cysteine
$H_2N-\overset{\overset{\displaystyle C_6H_5}{\overset{\displaystyle |}{\overset{\displaystyle CH_2}{|}}}}{^*CH}-COOH$ phenylalanine	$H_2N-\overset{\overset{\displaystyle CH_3\diagdown \diagup CH_3}{\overset{\displaystyle CH}{|}}}{^*CH}-COOH$ valine	
$H_2N-\overset{\overset{\displaystyle NH_2}{\overset{\displaystyle |}{\overset{\displaystyle (CH_2)_4}{|}}}}{^*CH}-COOH$ lysine	$H_2N-\overset{\overset{\displaystyle COOH}{\overset{\displaystyle |}{\overset{\displaystyle CH_2}{|}}}}{CH}-COOH$ aspartic acid	

Table 3 shows a few examples of α-amino acids present in proteins—note that with the exception of glycine, these compounds all possess an asymmetric carbon atom (marked with an asterisk) and so are capable of optical activity. It has also been shown that the α-amino acids isolated from proteins have the L-configuration, even though their optical rotation may be $(+)$ or $(-)$ (the reader should consult a good organic textbook if he is not familiar with this notation).

Hydrolysis of proteins also occurs naturally. Animals obtain α-amino acids from proteins in their diet via enzymic hydrolysis (enzymes are catalysts for the essential chemical reactions taking place in living organisms and consist of proteins which, in some cases, are combined with other compounds). These α-amino acids are then converted, under the influence of nucleic acids (see p. 83), to proteins required by the body. It should be pointed out at this stage that whilst some α-amino acids can be synthesised in the body, some cannot and therefore have to be obtained from protein in the diet. Because of this, these α-amino acids are described as *essential* α-amino acids (and those α-amino acids in Table 3 which are essential are underlined).

All plants need nitrogen so that they can produce proteins. Some plants can fix nitrogen directly from the atmosphere by means of bacteria in their root nodules, whilst most others take their nitrogen from the soil in the form of nitrate ions, etc.

The determination of protein structure is a formidable task. Since proteins are copolymers, the first problem is to determine which α-amino acid residues are present and the relative amounts of each. The protein can first be hydrolysed with mineral acid to yield a mixture of α-amino acids—this process can be represented as follows, where R, R' and R'' can represent different groups such as those shown in Table 3:

$$\text{\large\textsf{\char`\~}}NH.CHR.CO-NH.CHR'.CO-NH.CHR''.CO\text{\large\textsf{\char`\~}}$$

$$\downarrow \quad \text{addition of water}$$

$$H_2N-CHR-COOH + H_2N-CHR'-COOH$$
$$+ H_2N-CHR''-COOH \quad \text{and so on.}$$

This resultant mixture of α-amino acids can now be identified either by paper chromatography or by using an ion-exchange resin; the relative amounts are also determined.

The second problem is to find the sequence of the α-amino acid residues in the chain. This can be done using a selective method of chemical degradation, whereby the protein is broken down to smaller units in various ways. Identification of these smaller units then gives information concerning the original protein structure. For example, suppose that a protein contains four α-amino acids, denoted by the letters A, B, C and D. Suppose, also, that two methods of degradation give the following units:

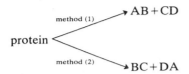

$$\text{protein} \begin{cases} \xrightarrow{\text{method (1)}} AB + CD \\ \xrightarrow{\text{method (2)}} BC + DA \end{cases}$$

From this information, the sequence of α-amino acid residues in the chain must be $-A-B-C-D-$. X-ray diffraction analysis is also a useful method for helping with this problem.

The third problem is to determine the shape of the protein molecule and again, X-ray diffraction has proved to be a powerful tool in doing this. For example, X-ray diffraction studies have shown that intramolecular hydrogen bonding between the $>N-H$ group of one residue and the $>C=O$ group of another can lead to coiling of the protein chain, the structure being described as an α-helix (see Figure 37(a)). This type of chain conformation is referred to as the *secondary* structure of the protein (whereas the chain composition and amino acid sequence described earlier is called the *primary* structure of the

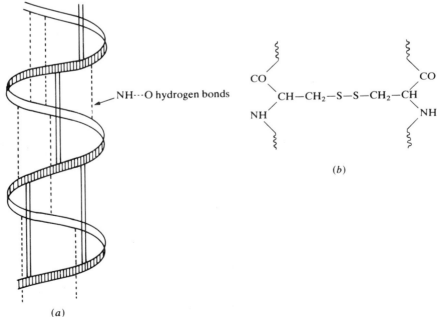

(a)

(b)

Figure 37 (a) Schematic representation, without structural detail, of the α-helix in a protein molecule, and (b) a disulphide link between two cysteine residues.

protein). Chains are sometimes also twisted, coiled or folded in space to give the overall molecular shape, called the *tertiary* structure, and this can arise because of various kinds of linkages between groups in different amino acid units which are sometimes quite widely separated along the chain. Clearly, hydrogen bonding will play a part in this. However, if the protein chain contains two or more cysteine residues (see Table 3), then disulphide linkages formed by oxidation of two —SH groups can result in large rings or loops being formed (see Figure 37(b)). Similarly, linkages formed by salt formation between a free amino group on one unit (from a lysine residue, for example—see Table 3) and a free carboxyl group on another (derived from aspartic acid, for example—see Table 3) can also hold folds or loops in the chain in position:

$$CH.CH_2.COOH + H_2N.(CH_2)_4.CH \rightarrow$$

$$CH.CH_2.COO^- H_3N^+.(CH_2)_4.CH$$

Proteins may be divided into *simple* and *conjugated* types, but we will not be considering conjugated proteins in this text. Simple proteins can be subdivided into *fibrous* and *globular* proteins. Fibrous proteins are insoluble in water, and are generally formed by helical chains packed or twisted round each other, as in bundles, giving them strength. Examples include collagen (in skin and in connective and other tissue), keratin (in hair and nails), and fibroin (in silk).

Globular proteins have some degree of solubility in water (either dissolving or forming colloidal solutions); the helical chains are in a much more coiled, compact or folded state compared to the fibrous proteins. Egg albumin is a globular protein.

Nucleic acids (or polynucleotides)

The synthesis of proteins within an animal is controlled by nucleic acids. Nucleic acids are polymers of *nucleotides*, where a nucleotide is a substance composed of a sugar, a phosphate group and a nitrogen base. In nucleic acids, the polymer chain consists of alternating sugar

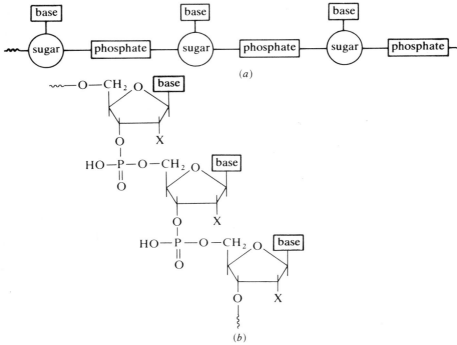

Figure 38 (a) General structure of a nucleic acid. In (b), the bonding between the sugar and the phosphate residues is shown. In DNA, X = H and in RNA, X = OH.

and phosphate groups, as shown in Figure 38. In *ribonucleic acid* (RNA), the sugar is ribose (Figure 39(a)) and the bases are cytosine and uracil (members of the pyrimidine group—see Figure 40(a)), and adenine and guanine (members of the purine group—see Figure 40(b))—these bases are illustrated in Figure 41. In *deoxyribonucleic acid* (DNA), the sugar is deoxyribose (Figure 39(b)) and the bases are the same as those in RNA except that thymine replaces uracil (Figure 41).

In DNA, two nucleic acid chains, each in the form of a helix, are coiled together and held by hydrogen bonds between specific pairs of bases on adjacent chains; an adenine unit (A) in one helix hydrogen bonds to a thymine unit (T) in the other, and similarly a guanine unit (G) hydrogen bonds to a cytosine unit (C) (see Figure 42(a)). Consequently, if the sequence of bases in one of the helices or strands of DNA is A—G—T—C, and so on, then the sequence in the other strand must be T—C—A—G, and so on. Figure 42(b) gives an example of base pairing by hydrogen bonding.

In cell division (or *mitosis*) the two helices of the DNA separate, each strand then acting as a 'template' for the synthesis of a new

HO—CH2 O
OH
H
H
H
H
OH OH
(a)
HO—CH2 O
OH
H
H
H
H
OH H
(b)

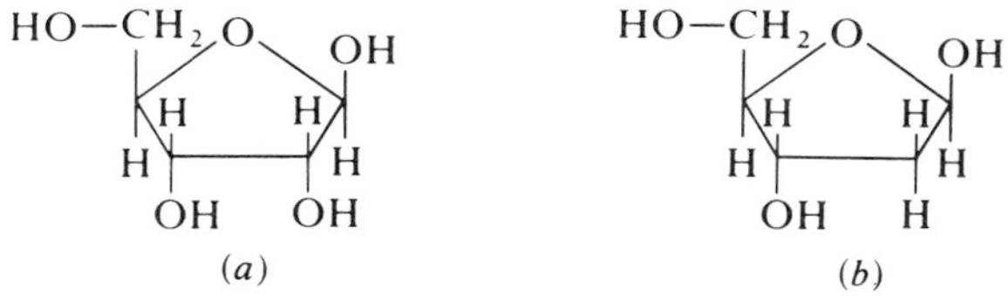

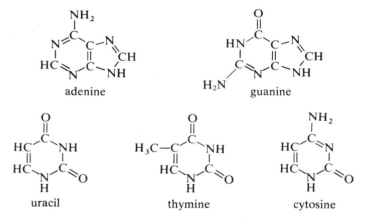

Figure 39 The structures of (a) ribose and (b) deoxyribose.

(a)
(b)

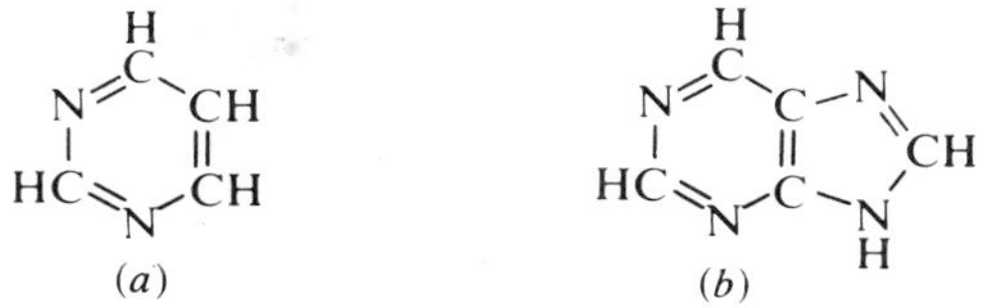

Figure 40 The structures of (a) pyrimidine and (b) purine.

adenine
guanine
uracil
thymine
cytosine

Figure 41 The structures of the bases present in DNA and RNA.

strand of DNA. Because base pairing is specific, the base sequence in one strand determines the base sequence in the complementary strand which is to be made. For example, strand X in Figure 42(a) will organise nucleotides in such a way (by base pairing) that the new strand which is formed will have structure Y, and hence the original Y will synthesise a new X in the same way. Hence, on cell division, the complete DNA molecule can be reproduced from just one of the strands.

In protein synthesis, constituent α-amino acids have to be assembled together in precisely the correct sequence. Two types of RNA molecules which are particularly important for ensuring that this occurs are *transfer* RNA and *messenger* RNA. In protein synthesis, therefore, a particular α-amino acid becomes attached to a

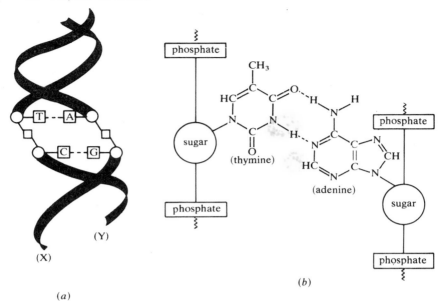

(a)

(b)

Figure 42 (a) Section of the double helix structure of DNA, where ○ and □ represent sugar and phosphate residues respectively, and the dotted lines represent hydrogen bonds. (b) Example of base pairing by hydrogen bonding.

particular transfer RNA. In turn, this transfer RNA becomes associated with a particular part of a messenger RNA molecule (this occurs by a process called *triplet coding*—a group of three adjacent bases in the transfer RNA fits on to the messenger RNA by the pairing of corresponding base 'partners'). If this process is then repeated, two adjacent α-amino acids can join together by a peptide link, and so on, to yield the complete protein. Note how this mechanism ensures that the α-amino acids are joined in the correct sequence. Since the messenger RNA molecules (which are single strands, unlike the DNA) are synthesised under the control of the DNA (again by base pairing), it is therefore the DNA which determines the protein structures.

Miscellaneous naturally occurring high polymers

The list of polymeric materials which could be included in this section is endless; we could discuss natural resins and polyesters, and many inorganic substances such as some oxides, silicates and complex phosphates—we could even include graphite.

However, one group which should be looked at briefly is the 'isoprene' group, which includes rubber, gutta percha and balata. The commercial use of rubber has a long history; Goodyear first vulcanised it in 1839, by heating with sulphur. Since then, a massive industry has been built up based on natural (and later, synthetic) rubber, due largely in the early years to the development of the motor industry. Natural rubbers are widely distributed and can be obtained from a variety of trees and plants.

The empirical formula for rubber is $C_{10}H_{16}$, and it is now accepted that rubber is a linear high polymer of methylbuta-1,3-diene (isoprene):

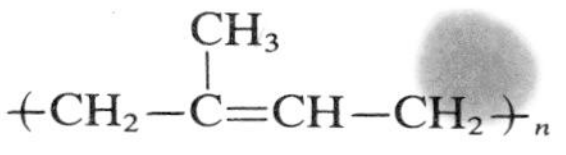

$$\begin{array}{c} CH_3 \\ | \\ +CH_2-C=CH-CH_2+_n \end{array}$$

However, as mentioned in Chapter 5, isomerism can occur, and it has been shown that rubber and gutta percha or balata are stereoisomers having *cis-* and *trans-* structures, respectively (Figure 43).

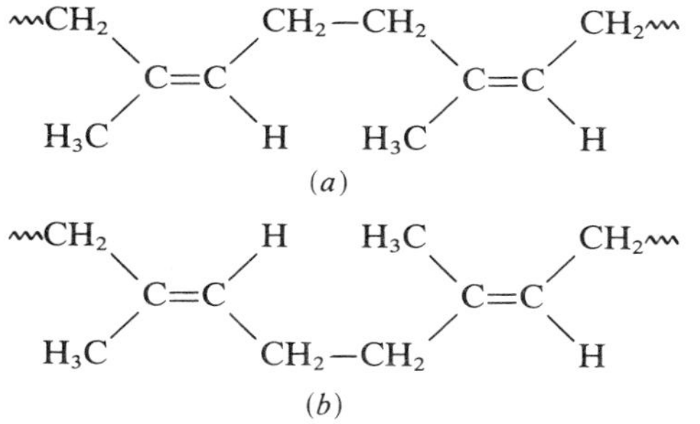

Figure 43 The structures of (a) rubber and (b) gutta percha and balata.

Natural rubber, when heated, becomes soft and tacky and will flow under its own weight; it will slowly dissolve in benzene. However, if rubber is vulcanised, i.e. heated with a small amount of sulphur (approximately 2 per cent), it becomes cross-linked and a dramatic change in properties is observed. Unvulcanised rubber exhibits 'creep' when stretched, i.e. progressive relaxation with time, since chains unravel and slip past each other. Vulcanised rubber, however, greatly resists creep and is much stronger when under tension than unvulcanised rubber. Solubility decreases with cross-linking and the vulcanised material may just swell when placed in a solvent. If rubber is vulcanised with a large proportion of sulphur (approximately 30 per cent), a very hard, chemically resistant material is produced known as ebonite or hard rubber; it is used for car battery cases.

Finally, it should be mentioned that the rate of reaction between natural rubber and sulphur can be substantially increased by the addition of 'accelerators', which consist of a wide range of organic compounds. Whilst the mechanism of vulcanisation is still not fully understood, the cross-links can be represented as follows:

$$\begin{array}{c} \text{\textasciitilde\textasciitilde\textasciitilde\textasciitilde\textasciitilde\textasciitilde\textasciitilde\textasciitilde\textasciitilde\textasciitilde\textasciitilde} \\ | \\ S_x \\ | \\ \text{\textasciitilde\textasciitilde\textasciitilde\textasciitilde\textasciitilde\textasciitilde\textasciitilde\textasciitilde\textasciitilde\textasciitilde\textasciitilde} \end{array}$$

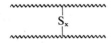

Practical Exercises and Further Reading 7

Practical exercises

Caution

Carry out these experiments in a well ventilated laboratory, or fume cupboard if possible.

1 Preparation of poly(methyl 2-methylpropenoate) (poly(methyl methacrylate) or 'Perspex'). To about 5 cm^3 of methyl 2-methyl-propenoate (methyl methacrylate) monomer in a test tube, add about 0.1 g of di(dodecanoyl) peroxide (lauroyl peroxide) initiator. Stopper the tube and gently upturn it several times to dissolve the initiator. After removing the stopper, clamp the tube in a beaker of water and maintain the temperature at about 85–90 °C. After 5–10 minutes the liquid becomes viscous, and after 10 minutes or so it sets to a hard polymer.
Given that the structure of di(dodecanoyl) peroxide is:

$$C_{11}H_{23}-\underset{\underset{O}{\|}}{C}-O-O-\underset{\underset{O}{\|}}{C}-C_{11}H_{23}$$

describe the function of this compound and write simple equations to illustrate how the polymer chain starts growing.

2 Preparation of poly(phenylethene) (polystyrene). To about 5 cm^3 of phenylethene (styrene) in a test tube, add about 0.1 g of di(dodecanoyl) peroxide initiator. Stopper the tube and gently upturn it several times to dissolve the initiator. After removing the stopper, clamp the tube in a beaker of water and maintain the temperature at 100 °C. After about 25 minutes, remove the tube from the water bath to cool, when the viscous liquid will begin to stiffen up. The polymer can be dissolved (if necessary) in a small amount of methyl-benzene (toluene). If this solution is now poured into a beaker containing a large excess of ethanol, poly(phenylethene) is precipitated as a white polymeric solid. How would you describe the function of the ethanol in this experiment?

3 Depolymerisation of a polymer. Place about 1.5 g of 'Perspex' chippings in a pyrex test tube fitted with a right-angled delivery tube, and allow the end of the delivery tube to dip into a receiving test tube which is immersed in a beaker of cold water. Heat the chippings gently with a bunsen and collect the distillate in the water-cooled tube. If the distillate is not colourless, it can be redistilled. The monomer you have collected can now be polymerised, if desired, as in experiment **1**. How does the thermal decomposition of poly(chloroethene) (polyvinyl chloride or PVC) differ from that of 'Perspex'?

4 Preparation of nylon 6,6. Place about 5 cm^3 of a 5% solution of hexanedioyl dichloride (adipyl chloride) in tetrachloromethane (carbon tetrachloride) in a small beaker. Add slowly, using a dropper, 5 cm^3 of a 5% aqueous solution of hexane-1,6-diamine (hexamethylenediamine) containing 5 drops of dilute (2 mol dm^{-3}) aqueous ammonia solution. This aqueous layer should rest on the tetrachloromethane layer—do *not* mix them. Using a pair of tweezers, carefully draw (from the centre of the beaker) a thread of nylon from the liquid interface—do not catch the thread on the sides of the beaker or the nylon thread may break. When the thread is long enough, it may be wrapped round a glass rod and the rod can then be rotated, acting like a pulley.

If the method is repeated but using decanedioyl dichloride (sebacoyl chloride) instead of hexanedioyl dichloride, nylon 6,10 is made. Write simple equations to illustrate how these polymers have been formed.

5 Drawing a nylon fibre. Warm carefully, using a bunsen, about 1 g of nylon 6 pellets in a small crucible. When the polymer has melted, draw a fibre from the melt using a thin wire or pin. Test the strength of the drawn fibre by hand. Outline the essential factors which make the nylons very good fibre-forming materials.

6 Preparation of a phenol–methanal (phenol–formaldehyde) resin. In a test tube, dissolve 1 g of phenol in 2 cm^3 of a 40 per cent aqueous solution of methanal. Add approximately 0.2 cm^3 of dilute (2 mol dm^{-3}) aqueous ammonia solution to the mixture (a small graduated syringe is suitable for doing this). Add some anti-bumping granules (this is important since the reaction is exothermic and severe bumping may occur on heating if the granules are not added), and after clamping the tube to a retort stand, heat the tube gently with a bunsen until the mixture turns milky white. Stop heating and the mixture should now separate into two layers—a bottom, yellow, viscous layer and a top white layer consisting mostly of water. Remove the top layer from the tube using a dropper or pipette, and heat the remaining bottom layer. The resin will turn darker yellow, will bubble, and will eventually set to a reddish-brown glass-like solid taking the shape of the tube. Describe how the various P/F ratios can dictate the outcome of these sorts of reaction.

7 Preparation of a urea–methanal resin. Place about 0.5 g of urea in a test tube and add approximately 1 cm^3 of concentrated hydrochloric acid and 5 cm^3 of water. After shaking the tube to dissolve the urea, add 10 drops of a 40 per cent aqueous solution of methanal. A white precipitate of urea–methanal resin will form. If no precipitate is observed, the mixture can be scratched with a glass rod which will bring about precipitation. Write down the first steps in the production of a linear polymer starting from urea and methanal. Why is it that this product, under suitable conditions, can now cross-link?

8 Preparation of a synthetic rubber (thiokol). In a beaker, dissolve 2 g of sodium hydroxide in about 50 cm^3 of water and gently boil the solution. Stir in 5 g of sulphur and when most of the sulphur has dissolved, allow the mixture to cool and then filter into another beaker. To the yellow-brown filtrate, add 10 cm^3 of 1,2-dichloroethane (ethylene dichloride) and keep the mixture at 70–80 °C, stirring continuously. Spongy, white lumps of rubber separate out which can be removed from the reaction mixture, washed well with water, and can be tested by hand for their rubber-like properties.

Outline the special requirements for a polymer to function well as an elastomer.

Suggestions for further reading

Billmeyer, F. W. (1971) *Textbook of Polymer Science* (2nd Edition), John Wiley.

Margerison, D. and East, G. C. (1967) *Introduction to Polymer Chemistry*, Pergamon Press.

Moore, W. R. (1963) *An Introduction to Polymer Chemistry*, University of London Press.

Parker, D. B. V. (1974) *Polymer Chemistry*, Applied Science Publishers.

Ray, N. H. (1978) *Inorganic Polymers*, Academic Press.

Saunders, K. J. (1973) *Organic Polymer Chemistry*, Chapman and Hall.

Tooley, P. (1971) *High Polymers*, John Murray.

Glossary

Addition polymerisation This type of polymerisation involves chain reactions, in which the chain carrier may be a free radical or an ion.

Amorphous regions These are regions of the polymer sample where the chains are tangled or in a state of disorder.

Chain molecule (*sometimes called polymer chain*) If the repetition of the repeating unit is linear (in much the same way as a chain is built up from its links), then the polymer molecules are often described as chain molecules or as polymer chains.

Cold-drawing This is the process of stretching a polymer such that its crystalline regions become aligned or oriented in the direction of the applied stress.

Cold-setting This is a polymer system which can be cured in the cold.

Condensation polymerisation This is a polymerisation process which proceeds in a stepwise manner by reaction between pairs of functional groups, accompanied by the loss of a small molecule at each stage of reaction. (N.B. Some polymerisations, such as the production of polyurethanes from diisocyanates and diols, can also be classed as condensation polymerisations even though no small molecules are eliminated.)

Copolymer This is a polymer made by polymerising two or more suitable monomers together.

Cross-linked polymer is formed when separate linear or branched chains are joined together along the chains by cross-links.

Crystalline regions Sometimes called 'crystallites', these are regions of the polymer sample where the chains are arranged in a highly ordered manner.

Curing This is the formation of extensive three-dimensional cross-linking in the final stages of production of a polymer.

Degree of polymerisation (*D.P.*) This is the number of repeating units in the polymer chain.

Depolymerisation This is the decomposition of a polymer by a stepwise loss of monomer units in a reaction which is essentially the reverse of polymerisation.

Elastomer Elastomers are amorphous polymers, above their glass transition temperatures at room temperature, and contain cross-links to prevent gross slipping of chains.

Extrusion This involves forcing the hot molten polymeric material through a shaped hole, or die, to form a rod, tube, sheet, ribbon, or other continuous length.

Fibre Polymers with symmetrical repeating units and high inter-chain forces can be used to make fibres, having high crystallinity and tensile strength.

Gel At some point in a polymerisation process (called the gel point), extremely large network structures may be formed, accompanied by a sudden change in polymer properties. For example, the reaction mixture changes from a viscous liquid to a jelly-like material which separates out of solution; this material is called a gel.

Glass transition Sometimes a polymer can change from a hard, brittle, glass-like substance to a soft, flexible, rubbery material as its temperature is raised. This is the glass transition, and the temperature at which this change in properties occurs is called the glass transition temperature (T_g). The glass transition is associated with the amorphous regions of the polymer.

Homopolymer A polymer whose repeating units are all of the same type or structure is called a homopolymer.

Inhibitor Inhibitors are substances which react with free radicals as soon as they are produced. They can, therefore, prevent free radical polymerisation from occurring until all the inhibitor has been used up.

Injection moulding This is when polymer, softened by heat, is injected under pressure into a cool mould where it hardens to give the required article.

Melting point of a polymer This is essentially the temperature (T_m) at which separation of chains in the crystalline regions of the polymer occurs, hence enabling the polymer to flow.

Melt spinning This involves pumping molten polymer at a constant rate under high pressure through a plate, called a spinneret, containing a number of holes. The liquid polymer 'jets' emerge downward from the face of the spinneret, usually into the air (where they solidify).

Monomer The monomer is the starting material from which the polymer is made.

Network During the production of a polymer, if cross-linking occurs to a high degree, a three-dimensional cross-linked or network polymer can be produced.

Number average relative molecular mass $(\bar{M}_n)$ This is one way of describing the average relative molecular mass of a polymer, and is defined as:

$$\bar{M}_n = \frac{\sum N_i M_i}{\sum N_i}$$

where N_i is the number of molecules with D.P. $= i$ and M_i is the relative molecular mass of molecules with D.P. $= i$.

Plastic Plastics are materials which have a lower degree of crystallinity than fibres, and can be softened or moulded at higher temperatures (their glass transition temperatures are above room temperature) unless they are highly cross-linked.

Plasticiser This is a substance which can be added to a polymer to reduce interchain forces and hence make movement of chain segments easier.

Polymer Sometimes called macromolecules, polymers are large molecules built up by the repetition of small, simple chemical units.

Polymerisation The process of forming a polymer is known as polymerisation.

Repeating unit The repeating unit of the polymer is usually equivalent, or nearly so, to the monomer, and forms the basis of the polymeric molecule.

Retarder Retarders are substances which compete with the monomer for free radicals during free radical polymerisation; hence, both the rate and degree of polymerisation are reduced. They are less reactive than inhibitors.

Thermoplastic Polymers which can be softened and moulded on heating are described as being thermoplastic.

Thermosetting This is a polymer system which can be cured by the application of heat.

Vulcanisation The process of introducing cross-links into the polymeric material is known as vulcanisation.

Weight average relative molecular mass $(\bar{M}_w)$ This is one way of describing the average relative molecular mass of a polymer, and is defined as:

$$\bar{M}_w = \frac{\sum w_i M_i}{\sum w_i}$$

where w_i is the total mass of all molecules whose D.P. $= i$ and M_i is the relative molecular mass of molecules with D.P. $= i$.

Index